Modeling, Simulation and Control of Wind Diesel Power Systems

Modeling, Simulation and Control of Wind Diesel Power Systems

Editor

Rafael Sebastián Fernández

MDPI • Basel • Beijing • Wuhan • Barcelona • Belgrade • Manchester • Tokyo • Cluj • Tianjin

Editor
Rafael Sebastián Fernández
Department of Electrical,
Electronic and Control
Engineering (DIEEC),
Universidad Nacional de
Educación a Distancia
(UNED)
Spain

Editorial Office
MDPI
St. Alban-Anlage 66
4052 Basel, Switzerland

This is a reprint of articles from the Special Issue published online in the open access journal *Energies* (ISSN 1996-1073) (available at: https://www.mdpi.com/journal/energies/special_issues/Modeling_Simulation_Control_Wind_Diesel_Power_System).

For citation purposes, cite each article independently as indicated on the article page online and as indicated below:

LastName, A.A.; LastName, B.B.; LastName, C.C. Article Title. *Journal Name* **Year**, *Volume Number*, Page Range.

ISBN 978-3-0365-3831-0 (Hbk)
ISBN 978-3-0365-3832-7 (PDF)

Contents

About the Editor

Rafael Sebastián Fernández received an M.Sc. degree in industrial engineering from the Universidad Politécnica de Madrid, Madrid, Spain, in 1983, and a Ph.D. degree in electrical engineering from the Spanish University for Distance Education (UNED), Madrid, in 2003. He was an Assistant Professor with UNED from 1984 to 1989. From 1990 to 2000, he worked as an Electronic and Control Design Engineer in several companies in Spain. In 2001, he returned to UNED, where he is currently working with the Electrical, Electronic, and Control Engineering department. He is an Associate Professor of electronics technology, lecturing on analog electronics, electronic instrumentation, and distributed control systems, among other subjects. He has authored 60 papers published in international conferences and journals.

His research interest areas are the modeling, simulation, and control of isolated microgrids, based on wind turbine generators alone or combined with diesel generators or hydro turbine generators. Additionally, his research field embraces energy storage systems mainly based on batteries and flyweels.

Preface to "Modeling, Simulation and Control of Wind Diesel Power Systems"

Welcome to the Special Issue: "Modeling, Simulation and Control of Wind Diesel Power Systems (WDPS)". Seven articles have been published in this Special Issue. The article that best follows the "spirit" of the Special Issue is the "Review on Dynamic Simulation of Wind Diesel Isolated Microgrids". I recommend that potential readers start with this article. Two more articles deal strictly with WDPS: "Flywheel Energy Storage and Dump Load to Control the Active Power Excess in a Wind Diesel Power System" presents dynamic simulations of a WDPS that includes a flywheel energy storage system (FESS) and shows how the FESS stabilizes the microgrid and improves the system power quality. "Study of the Intelligent Control and Modes of the Arctic-Adopted Wind–Diesel Hybrid System" shows the architecture of an advanced control system with different control options for WDPS located in the artic area, and how to tackle the very low temperatures and the icing of the blades of the wind turbine generators (WTGs). Another two articles also present dynamic simulations of isolated microgrids: "Study and Simulation of a Wind Hydro Isolated Microgrid" deals with modeling a high-penetration wind-hydro microgrid, and its main simulation is a transition from the only wind operation to the wind-hydro operation. "An Adaptive Control Scheme for Variable Speed Wind Turbines Providing Frequency Regulation in Isolated Power Systems with Thermal Generation" presents adaptative control that commands variable-speed WTGs to support frequency in an isolated system with controlled thermal generators, minimizing the WTGs mechanical and electrical efforts. Last but not least, two articles show logistic simulations: "Long-Range Integrated Development Analysis: The Cuban Isla de la Juventud Study Case" studies 25% and 100% scenarios of renewable energy penetration in the Cuban island of "la Juventud". "Taking into Consideration the Inclusion of Wind Generation in Hybrid Microgrids: A Methodology and a Case Study" presents methodology to facilitate the addition of wind power generation to photovoltaic-diesel isolated microgrids.

I hope that this Special Issue is helpful to students, researchers, and professors.

Rafael Sebastián Fernández
Editor

energies

Editorial

Modeling, Simulation and Control of Wind Diesel Power Systems

Rafael Sebastián

Department of Electrical, Electronic and Control Engineering (DIEEC), Universidad Nacional de Educación a Distancia (UNED), 28040 Madrid, Spain; rsebastian@ieec.uned.es

Wind diesel power systems (WDPS) are isolated microgrids which combine diesel generators (DGs) with wind turbine generators (WTGs). WDPS are on many cases the result of adding WTGs to a previous existing diesel power plant located in a remote place where there is an available wind resource. By means of the WTGs supplied power, the fuel consumption and the CO_2 emissions are reduced [1]. WDPSs are isolated power systems with low inertia where important system frequency and voltage variations occur [2]. WDPS dynamic modeling and simulation allows short-term simulations to obtain detailed electrical variables transients, so that the WDPS stability and power quality can be tested. The main subject of this Special Issue in the *Energies* journal is about WDPS dynamic modelling and simulation. The topics included in the main subject were: the simulation and control of different architectures of WDPS, the dynamic modelling of the different WDPS components, the consideration of the different modes of operation of WDPS: diesel only, wind–diesel [3] and wind only [4]; the advantages and drawbacks of using WTG of fixed speed [5] or variable speed [6] types in WDPS, and the use of energy storage systems (ESS) based on batteries [7], flywheel [8], ultracapacitors [9], etc., and the benefits that ESSs provide to the WDPS power quality, stability and reliability [10]. Additionally, the related topics of logistic simulation of WDPS or studies about the sizing of the WDPS different components are also included. WDPS are isolated microgrids, so the dynamic simulation of related AC isolated microgrids such as those with hydro-turbines [11], photovoltaics, ship microgrids [12], etc., can be also part of this Special Issue.

The Special Issue was closed on 30 April 2021 with seven articles published on it. The article that most follows the "spirit" of the Special Issue is "Review on Dynamic Simulation of Wind Diesel Isolated Microgrids" [1]. I would recommend the potential readers of this Special Issue to start with this article. Two more articles deal strictly with WDPS: "Flywheel Energy Storage and Dump Load to Control the Active Power Excess in a Wind Diesel Power System" [8] presents dynamic simulations of a WDPS which includes a flywheel energy storage system (FESS) and shows how the FESS stabilizes the microgrid and improves the system power quality, and "Study of the Intelligent Control and Modes of the Arctic-Adopted Wind–Diesel Hybrid System" [13] shows the architecture of an advanced control system with different control options for WDPS located in the artic area, and how to tackle the very low temperatures and the icing of the blades of the WTGs. Another two articles also present dynamic simulations of isolated microgrids: "Study and Simulation of a Wind Hydro Isolated Microgrid" [11] deals with the modeling of a high penetration wind–hydro microgrid and its main simulation is a transition from the wind only operation to the wind–hydro operation, and "An Adaptive Control Scheme for Variable Speed Wind Turbines Providing Frequency Regulation in Isolated Power Systems with Thermal Generation" [14] presents an adaptative control that commands variable speed WTGs to support frequency in an isolated system with controlled thermal generators, minimizing the WTGs mechanical and electrical efforts. Finally, two articles show logistic simulations: "Long-Range Integrated Development Analysis: The Cuban Isla de la Juventud Study Case" [15] studies the 25% and 100% scenarios of renewable

Citation: Sebastián, R. Modeling, Simulation and Control of Wind Diesel Power Systems. *Energies* **2022**, *15*, 1712. https://doi.org/10.3390/en15051712

Received: 8 February 2022
Accepted: 17 February 2022
Published: 25 February 2022

energy penetration in the Cuban island of "la Juventud", and "Taking into Consideration the Inclusion of Wind Generation in Hybrid Microgrids: A Methodology and a Case Study" [16] presents a methodology to facilitate the addition of wind power generation to photovoltaic–diesel isolated microgrids.

I hope that this Special Issue be helpful to students, researchers, and professors.

Funding: This research received no external funding.

Institutional Review Board Statement: Not applicable.

Informed Consent Statement: Not applicable.

Conflicts of Interest: The author declares no conflict of interest.

References

1. Sebastián, R. Review on Dynamic Simulation of Wind Diesel Isolated Microgrids. *Energies* **2021**, *14*, 1812. [CrossRef]
2. Sebastián, R. Smooth transition from wind only to wind diesel mode in an autonomous wind diesel system with a battery-based energy storage system. *Renew. Energy* **2008**, *33*, 454–466. [CrossRef]
3. Sebastián, R. Reverse power management in a wind diesel system with a battery energy storage. *Int. J. Electr. Power Energy Syst.* **2013**, *44*, 160–167. [CrossRef]
4. Sebastián, R.; Quesada, J. Distributed control system for frequency control in a isolated wind system. *Renew. Energy* **2006**, *31*, 285–305. [CrossRef]
5. Sebastian, R. Modelling and simulation of a high penetration wind diesel system with battery energy storage. *Int. J. Electr. Power Energy Syst.* **2011**, *33*, 767–774. [CrossRef]
6. Lukasievicz, T.; Oliveira, R.; Torrico, C. A Control Approach and Supplementary Controllers for a Stand-Alone System with Predominance of Wind Generation. *Energies* **2018**, *11*, 411. [CrossRef]
7. Sebastian, R.; Pena-Alzola, R. Study and Simulation of a Battery Based Energy Storage System for Wind Diesel Hybrid Systems. In Proceedings of the 2012 IEEE International Energy Conference and Exhibition (ENERGYCON), Florence, Italy, 9–12 September 2012; pp. 563–568.
8. Sebastián, R.; Peña-Alzola, R. Flywheel Energy Storage and Dump Load to Control the Active Power Excess in a Wind Diesel Power System. *Energies* **2020**, *13*, 2029. [CrossRef]
9. Tarkeshwar, M.; Mukherjee, V. Quasi-oppositional harmony search algorithm and fuzzy logic controller for load frequency stabilisation of an isolated hybrid power system. *IET Gener. Transm. Distrib.* **2015**, *9*, 427–444. [CrossRef]
10. Sebastián, R. Battery energy storage for increasing stability and reliability of an isolated Wind Diesel power system. *IET Renew. Power Gener.* **2017**, *11*, 296–303. [CrossRef]
11. Sebastián, R.; Nevado, A. Study and Simulation of a Wind Hydro Isolated Microgrid. *Energies* **2020**, *13*, 5937. [CrossRef]
12. Bø, T.I.; Johansen, T.A. Battery Power Smoothing Control in a Marine Electric Power Plant Using Nonlinear Model Predictive Control. *IEEE Trans. Control Syst. Technol.* **2017**, *25*, 1449–1456. [CrossRef]
13. Elistratov, V.; Konishchev, M.; Denisov, R.; Bogun, I.; Grönman, A.; Turunen-Saaresti, T.; Lugo, A.J. Study of the Intelligent Control and Modes of the Arctic-Adopted Wind–Diesel Hybrid System. *Energies* **2021**, *14*, 4188. [CrossRef]
14. Fernández-Guillamón, A.; Martínez-Lucas, G.; Molina-García, Á.; Sarasua, J.I. An Adaptive Control Scheme for Variable Speed Wind Turbines Providing Frequency Regulation in Isolated Power Systems with Thermal Generation. *Energies* **2020**, *13*, 3369. [CrossRef]
15. Alberto Alvarez, E.; Korkeakoski, M.; Santos Fuentefría, A.; Lourdes Filgueiras Sainz de Rozas, M.; Arcila Padura, R.; Luukkanen, J. Long-Range Integrated Development Analysis: The Cuban Isla de la Juventud Study Case. *Energies* **2021**, *14*, 2865. [CrossRef]
16. Arribas, L.; Bitenc, N.; Benech, A. Taking into Consideration the Inclusion of Wind Generation in Hybrid Microgrids: A Methodology and a Case Study. *Energies* **2021**, *14*, 4082. [CrossRef]

energies

MDPI

Article

Study of the Intelligent Control and Modes of the Arctic-Adopted Wind–Diesel Hybrid System

Viktor Elistratov [1], Mikhail Konishchev [1], Roman Denisov [1], Inna Bogun [1,*], Aki Grönman [2], Teemu Turunen-Saaresti [2] and Afonso Julian Lugo [2]

1 Higher School of Hydraulic and Power Engineering Construction, Peter the Great St. Petersburg Polytechnic University (SPbPU), Polytechnicheskaya, 29, 195251 St. Petersburg, Russia; elistratov@spbstu.ru (V.E.); mak@vtr-engineering.ru (M.K.); denisov_rs@spbstu.ru (R.D.)

2 School of Energy Systems, LUT University, FI-53851 Lappeenranta, Finland; aki.gronman@lut.fi (A.G.); teemu.turunen-saaresti@lut.fi (T.T.-S.); lugoafonso@gmail.com (A.J.L.)

* Correspondence: bogun_iv@spbstu.ru

Abstract: For energy supply in the Arctic regions, hybrid systems should be designed and equipped to ensure a high level of renewable energy penetration. Energy systems located in remote Arctic areas may experience many peculiar challenges, for example, due to the limited transport options throughout the year and the lack of qualified on-site maintenance specialists. Reliable operation of such systems in harsh climatic conditions requires not only a standard control system but also an advanced system based on predictions concerning weather, wind, and ice accretion on the blades. To satisfy these requirements, the current work presents an advanced intelligent automatic control system. In the developed control system, the transformation, control, and distribution of energy are based on dynamic power redistribution, dynamic control of dump loads, and a bi-directional current transducer. The article shows the architecture of the advanced control system, presents the results of field studies under the standard control approach, and models the performance of the system under different operating modes. Additionally, the effect of using turbine control to reduce the effects of icing is examined. It is shown that the advanced control approach can reduce fuel consumption in field tests by 22%. Moreover, the proposed turbine control scheme has the potential to reduce icing effects by 2% to 5%.

Keywords: wind energy; hybrid systems; harsh climatic; pitch-control; intelligent control system; icing prediction; predictive analytics; adapted technologies

Citation: Elistratov, V.; Konishchev, M.; Denisov, R.; Bogun, I.; Grönman, A.; Turunen-Saaresti, T.; Lugo, A.J. Study of the Intelligent Control and Modes of the Arctic-Adopted Wind–Diesel Hybrid System. *Energies* **2021**, *14*, 4188. https://doi.org/10.3390/en14144188

Academic Editor: Rafael Sebastián Fernández

Received: 1 May 2021
Accepted: 7 July 2021
Published: 11 July 2021

1. Introduction

Most of the Russian and Finnish territories are located in a cold climate. Almost the entire northern part of these countries, as well as Central Siberia and Yakutia in Russia, fall within the cold polar zone of the Arctic, a zone of extremely low temperatures. In these areas, the duration of winter significantly exceeds that of summer with temperatures dipping close to −50 °C. Conditions here are unique, yet approximately 2.5 million people live in these areas. This is more than the total number of people residing in the Arctic areas of the seven other Arctic nations, all of which have less severe climatic conditions [1].

A significant part of the Arctic territory belongs to the decentralized energy supply zone. This zone is characterized by weak infrastructure associated with its remoteness from regional centers, and electricity is mainly produced by diesel power plants operating on expensive imported fuel. In the Russian areas of this zone, there are approximately 900 diesel power plants in operation, which produce an energy output of about 3.0 billion kWh annually [1]. The main challenges of supplying power to isolated consumers are the high logistical costs associated with the delivery of fuel and equipment for diesel power plants, the limited transport infrastructure, and, consequently, the high cost of fuel. Additionally, the operating costs of diesel power plants and specific fuel consumption are high, and there is typically no monitoring or control automation. As a result, the levelized cost of electricity (LCOE) at diesel power plants in the Arctic regions is 0.25–2.0 Euro/kWh, which is much more expensive than in the centralized energy supply zones. To illustrate, the International Energy Agency estimates LCOE values of below 0.1 Euro/kWh in Europe for coal, gas, and nuclear energy [2]. In addition, significant damage to the vulnerable natural environment of the Arctic is being caused by the emissions from fuel combustion products (40 million tons of carbon dioxide (CO_2), 80 thousand tons of sulfur oxides (SO_x), 600 thousand tons of nitrogen oxides (NO_x) annually), as well as by the storage of fuel barrels [3].

The Far North and the Arctic regions are characterized by high wind energy potential. In general, Russian and Finnish technical wind energy resources are 10^{16} kWh/year, about 16% of which are concentrated in the European and Asian northern coastal zones, creating a solid foundation for their effective use [4]. Given this high wind energy potential, optimization and modernization of existing power supply systems and the construction of new ones can be effectively carried out based on modular wind–diesel–storage hybrid systems adapted to Arctic conditions. These systems offer a good opportunity to overcome the challenges related to the intermittent and fluctuating nature of wind energy production, and they can also reduce the use of fossil fuels in distributed energy generation [5,6]. The application of wind–diesel hybrid systems will reduce the use of diesel fuel by 10–60% and increase the standard service life of diesel generator sets by two to three times [7–10]. In terms of CO_2 emissions, a study by Kazem et al. predicts an over 20% reduction in emissions when using a hybrid wind turbine–diesel engine system as compared with a diesel-only system [3]. It is worth mentioning that, in addition to the time-shifting role of the batteries in these systems, they can also improve the reliability and power quality as highlighted, for example, by Ansari and Velusami [11].

However, one of the key challenges to the optimal operation of Arctic wind turbines is icing. The impact of icing on wind turbine performance can be significant during the cold winter months. Turkia et al. [11] predict an approximately 17% reduction in turbine performance below nominal power due to icing, and Wei et al. [12] report power generation losses as high as 30%. Different passive and active anti-icing systems have been considered to reduce these effects, including special coatings, black paint, and heating [12]. One option that has not received much interest is airfoil pitch control. Nonetheless, several studies have shown that the optimal airfoil angle of attack can be affected by icing, for example, [13,14], introducing the potential to control the pitch differently during normal and icing operation modes.

Proportional integral (PI) control has traditionally been used in hybrid energy systems [9,15], but it poses problems regarding frequency regulation [11]. Increasing system stability has also been investigated recently [16]. The existing options to overcome the challenges described include, for example, the use of genetic algorithms [11] and fuzzy logic [17]. One potential approach to the control of a hybrid system is to use weather forecasts to help optimize the energy system's efficiency [18–20]. Under this approach, typically, wind forecast data are used to control the system, but icing, for example, is not considered although it can have a significant effect on performance especially in Arctic environments.

According to Elistratov et al. [4], wind–diesel hybrid systems should consider the following design constraints:

- Limitations in equipment and fuel delivery due to the short periods in which transportation is possible;
- The need for quick installation and construction without the use of heavy lifting and transport equipment in the absence of roads;
- The possibility for maintenance without the involvement of qualified specialists.

Additionally, the hybrid system must have a high degree of automation, including adaptive algorithms and intelligent control, and a remote monitoring and diagnostic system to optimize expensive diesel fuel usage.

Given the above, this study develops an existing field operating control system to improve its autonomous operation in Arctic conditions and maximize diesel fuel savings. It is shown that carbon-neutral technologies can be highly effective in Arctic zones subject to advanced control and reliable and safe operation. The study aims to identify how the system control approach affects system performance. Two control schemes are studied and compared with the use of a diesel engine only: 1. load following mode, and 2. cycle charge with short-term forecasting including icing effects. Additionally, a new wind turbine control method is proposed to decrease the effects of icing. The key novelties of the study are:

- The presentation of advanced control that includes climatic forecasting (wind speed, icing, etc.);
- The presentation of a new control approach to limit the effects of wind turbine icing.

The article is structured as follows. First, the hybrid system's control methods are explained. Next, the icing modeling and the novel pitch control approach are presented. In the Results section, before the discussion and conclusions, the effects of the different control methods are examined. This analysis is followed by the demonstration of the potential of a combined pitch and tip-to-speed ratio control approach to reduce the effects of icing.

2. Materials and Methods

This section presents the general layout of the proposed intelligent automatic control system (IACS) (Figure 1). Subsequently, the two parts of IACS, namely, standard control and advanced control, are explained in detail. This section also presents a novel turbine control scheme for reducing the effects of icing.

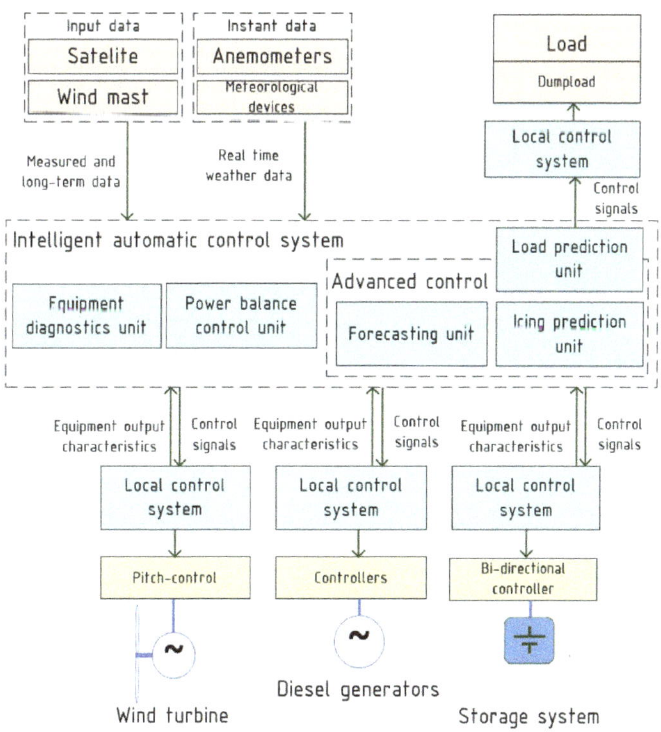

Figure 1. Functional diagram of IACS.

To solve the challenges related to the hybrid energy system's fuel economy, a control methodology has been proposed by Elistratov et al. [21]. Their study argues that an intelligent automatic control system must [21]:

1. In real time, maximize the energy output of the wind power plant and fuel economy while covering the required load;
2. Provide remote monitoring of the hybrid system's parameters and operating modes;
3. Provide intelligent dispatching of the equipment, ensuring the maximum degree of autonomy;
4. Monitor the condition of the equipment, analyze the statistics of wind–diesel operating modes, and provide forecasting of the wind regime;
5. Ensure scheduling of equipment operation, maintenance, risk assessment, and emergency prevention interventions;
6. Duplicate the main controller of the system and the control and measuring systems; in an emergency, the possibility of manual control should be provided;
7. Be adaptable and supply energy around the clock, including in the event of a failure of the generating equipment.

Structurally, IACS consists of the following units, which are presented in Figure 1 and explained in the following two sub-sections:

- An equipment diagnostics unit (supervisory control and data acquisition of each system element);
- A power balance control unit that distributes energy between the system's generating equipment;
- A forecasting unit;
- An icing prediction unit.

The role of the first two units is to allow the system to achieve high renewable energy penetration. In contrast, the last two units compose an additional advanced control system that allows operation in harsh climatic conditions.

2.1. Power Balance Control and Equipment Diagnostics Units (Standard Control)

IACS is the software part of the "conversion, control and energy distribution" module of hybrid energy systems. This module provides the possibility of maximizing energy production from renewable energy sources due to the dynamic redistribution of power between the elements of the hybrid system and, as a result, minimizing fuel consumption with the option to disconnect the diesel generator entirely when renewable energy sources have sufficient capacity.

Figure 2 presents the hardware part of this module [22]. The hardware consists of two power devices for dynamic power balance control (i.e., the bi-directional current transducer and controlled dump load) and the main controller, both of which perform high-level control. The energy sources of the autonomous hybrid system are divided into two categories: leading and following sources.

Leading sources can be either the diesel component (as the main source, defining a supply voltage) or the bi-directional current transducer with connected batteries (in autonomous inverter mode) while the following sources adapt to the main source's voltage and generate power to the grid (e.g., the wind component).

If the capacity of the diesel and wind components, averaged over a certain period, exceeds the total power consumption, then to achieve the maximum use of renewable energy and thus to maximize diesel fuel economization, it is possible to turn off all diesel generators. In this case, the leading source becomes the bi-directional current transducer, which goes into standalone inverter mode and generates a network voltage.

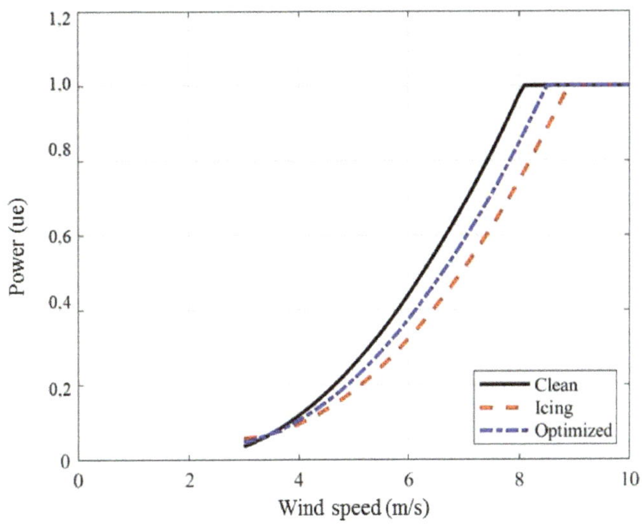

Figure 2. The "conversion, control and energy distribution" module of hybrid energy systems. K1: dump load controller; K3: common module controller; K2: bi-directional current transducer controller; 1.1: control stage current; 1.2: control circuits and internal power control circuits; 1.3: control and measurement of output electrical parameters; 2.1: control and measurement of output electrical parameters; 2.2: control and management circuit of the internal power circuit of the bi-directional current transducer; 2.3: control and measurement of electrical parameters on the battery side.

2.2. Forecasting and Icing Prediction Units (Advanced Control)

The intermittent and fluctuating nature of wind energy production increases the importance of short-term weather forecasting in energy systems. With renewables being introduced into isolated power grids, the inherent uncertainty associated with weather forecasts places significant strain on existing off-grid power systems. These challenges lead to power quality and stability issues and affect both power grid management and balancing. Moreover, efficient system control requires accurate estimations of both energy supply and demand, which further highlights the importance of weather forecasting. In general, energy demand is more stable than renewable energy production, which is directly influenced by local weather systems. It is, however, important to acknowledge that unexpected peaks in demand can occur, for example, due to extreme weather. Poor weather predictions can lead to various problems in off-grid systems with detrimental economic and environmental effects. These challenges include the possibility of power shortages, the need for additional spinning or non-spinning reserves, and the increased use of diesel fuel. Another possible scenario is that the system can produce a large oversupply of energy, whereby diesel fuel will be burned needlessly. These considerations fully justify the need for high-quality weather predictions covering 10 to 60-min time spans to ensure efficient grid supply and demand balancing [23].

Among traditional short-term forecasting methods such as the Auto-Regressive Integrated Moving Average (ARIMA), many modern processes use a form of deep learning known as recurrent neural networks (RNNs). A popular type of RNN, which is applied here, is the Long Short-Term Memory (LSTM) network. The models predicting wind characteristics and power output considered in this article are:

- Integrated autoregressive models (ARIMA) [24,25];
- Probabilistic models (Markov chain) [26,27];
- Statistical methods (GAMLSS);
- Machine learning (XGBoost, Random forest);
- Neural networks [28,29];

- Combined systems [30].

The ARIMA models are the most commonly used class of models for stationary signal forecasting (or a signal that can be made stationary). These models support random walk, seasonal trend, non-seasonal exponential smoothing, and autoregressive models.

Lags of the stationarized series in the forecasting equation are called "autoregressive" terms, while "moving average" terms describe the lags of the forecast errors. A time series, which needs to be differenced to be made stationary, is said to be an "integrated" version of a stationary series. Random-walk and random-trend models, autoregressive models, and exponential smoothing models are all considered special cases of ARIMA models.

The learning and encoding of signal temporal features are enabled by the RNN. This is an ideal approach to forecast signals, which are reasonably predictable based on past events. LSTM networks are recurrent networks that can overcome some of the historic challenges related to the training of recurrent networks, such as the vanishing gradients problem. This study will not go into the detail of evaluating and comparing forecasting models and will adopt the LSTM model due to its universally accepted ability to predict wind speed and load and perform predictive diagnostics of equipment condition.

Wind power plant output forecasts are based both on weather conditions and the power curves of the turbines. Moreover, at least one numerical weather forecast model should be integrated into the model being developed. These weather models will help to predict global weather patterns and their effects on local conditions. The numerical weather model (NWM) used to consider information other than data from local station observations is the NEMS4 model [31]. This model is provided free of charge by MeteoBlue (meteoblue AG, Basel, Switzerland) for a given date range and station. The data are provided in raw format.

The weather data prediction unit is connected to the icing prediction unit. When operating a wind turbine in a cold climate, additional power losses occur due to several types of icing: heavy frosting of the blades (in temperatures below −25 °C), sedimentary (cloudy) icing, and atmospheric icing (Figure 3).

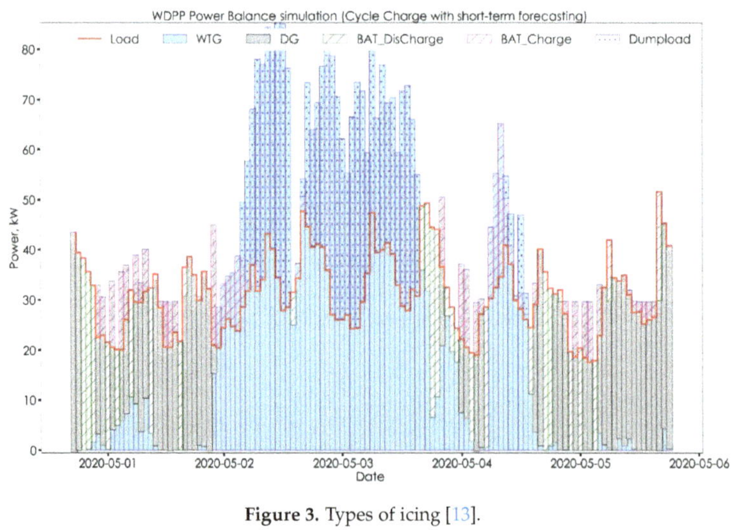

Figure 3. Types of icing [13].

Modern wind turbines include proven technical solutions to enable their operation in temperatures as low as −35 °C [32]. However, it is not only temperature that is important but also the duration of icing. For areas with long winter seasons, it is important to strengthen the control system by adding an icing prediction unit. This will make it possible to effectively use the existing systems to protect against ice, which can grow intensively on the surface of the blades.

According to the Makkonen theory [33], the functionality of the icing intensity indicator depends on the predicted weather parameters. Based on these calculations, a decision is made to turn on the protection system. The intensity of icing is determined by the following equation:

$$I = \alpha \times \beta \times \gamma \times LWC \times v \times A, \quad (1)$$

where α is the collision efficiency factor; β is the coefficient of sticking efficiency; γ is the coefficient of efficiency of growth (accretion); LWC is the liquid water content in the air (mass particle concentration), kg/m^3; v is the speed of incoming airflow (particle velocity), m/s; A is the cross-sectional area of the wind turbine blade (relative to the direction of the airflow velocity vector), m^2. LWC values and alpha coefficients depend on the weather parameters (pressure, temperature, humidity, specific water content in the environment, etc.). In this article, the prediction of the onset of atmospheric icing is based on the occurrence of the conditions presented in Table 1.

Table 1. Conditions for atmospheric icing (adopted from [34]).

Parameter	Condition
Wind Speed	>3 m/s
Temperature	−4 °C > T > −20 °C
Relative Humidity	>95%

To protect the blades from ice, special anti-icing and de-icing systems are used, as described in detail in [32,35]. In the icing prediction unit, the input data are acquired from meteorological instruments (weather data), wind measuring systems (wind speed, data correlation for the "heated–unheated anemometer" system), and directly from the wind turbine (power). When the output power from the wind turbine drops and there are conditions for atmospheric icing (Table 1), the system produces a signal to turn on the anti-icing system.

All icing protection systems are divided into two types: active and passive. Active systems require additional power from their own system (these include all anti-icing systems installed inside or outside the blade). Passive systems do not incur additional costs when operating the wind turbine (de-icing systems, for example, painting the blades in black). The pitch-control system for wind turbines with a capacity of more than 1 MW is categorized as a passive system since it is preinstalled and does not incur substantial additional costs to be operated. However, for wind turbines of a maximum 1 MW capacity, a feasibility study concerning the application of the regulation system is required.

In the case of a pitch-control in a lower capacity turbine, it is necessary to compare cost and effectiveness. The possible effect can be estimated based on the power output increase. Figure 4 shows how the turbine power coefficient changes due to icing based on the airfoil data reported by Homola et al. [36]. The calculations are based on Wilson's equation [37] and involve different angles of attack and tip-to-speed ratios. It is noticeable that, when icing occurs, the changes in the ratio of the lift and drag forces influence performance but so do changes in the tip-to-speed ratio.

In the development of a pitch-control-based approach, optimal drag-to-lift ratios from several references including airfoil performance data for both clean and icing conditions [32,33,35–39] are used. The data used include several wind speeds, as seen in Table 2. Wilson's equation [37] with different angles of attack and tip-to-speed ratios is used to predict the maximum power coefficient of a wind turbine. The results under four conditions are presented: 1. clean turbine, 2. turbine under icing conditions, 3. turbine under icing conditions with pitch control, and 4. turbine under icing conditions with combined pitch and tip-to-speed ratio control. The results for different wind speeds are summarized in Table 2. This analysis reveals that the pitch control can overcome some of the icing effects, but the combined pitch and tip-to-speed ratio control has even higher loss reduction potential. Therefore, the potential of the combined pitch and tip-to-speed ratio control is examined more closely in the following section.

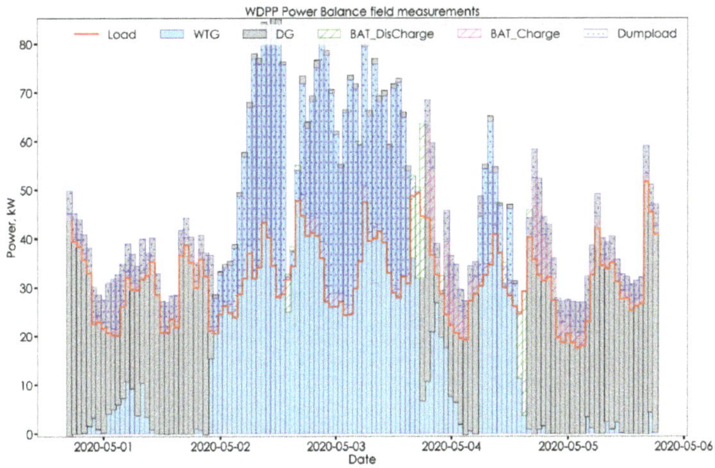

Figure 4. Effects of tip-to-speed ratio and angle of attack on turbine power coefficient for (**a**) clean turbine and (**b**) during icing based on the drag to lift data (adopted from [34]).

Table 2. Results summary of the performance analysis for a wind turbine with and without icing (adopted from [34]). The results are presented for a clean turbine (Cp_{clean}), a turbine under icing conditions (Cp_{icing}), a turbine under icing conditions with pitch control (Cp_{α}), and a turbine under icing conditions with pitch and tip-to-speed ratio control ($Cp_{\alpha\lambda}$).

Wind Speed (m/s)	Cp_{clean}	λ	α (°)	Cp_{icing}	$Cp_{\alpha\lambda}$	λ_{opt}	Cp_{α}	α_{opt} (°)
2.2	0.4755	5.6	5	0.4157	0.4526	4.6	0.4499	0
3.2	0.4565	4.7	4	0.3495	0.3848	2.8	0.3597	9
6	0.5126	8.3	5	0.3821	0.4340	3.9	0.3907	6
7.3	0.4984	7.0	6	0.3461	0.4078	3.2	0.3564	5
19.2	0.4173	3.5	2	0.3829	0.4002	3.1	0.3989	5

3. Results

This section presents the results of two hybrid system control modes and compares their performance with that of the diesel-only system. The data are presented for a conventional load following mode based on field data and for a simulated IACS operating mode. Additionally, the potential of the combined pitch and tip-to-speed ratio control approach is demonstrated, and the turbine icing modeling approach is verified.

3.1. Hybrid System Modes

A comparative analysis of the operating modes of an existing Arctic wind–diesel hybrid system was carried out to assess the practical benefits of the implementation of intelligent control algorithms. The correlation between the structure of the considered equipment and high renewable energy penetration is illustrated in Figure 5.

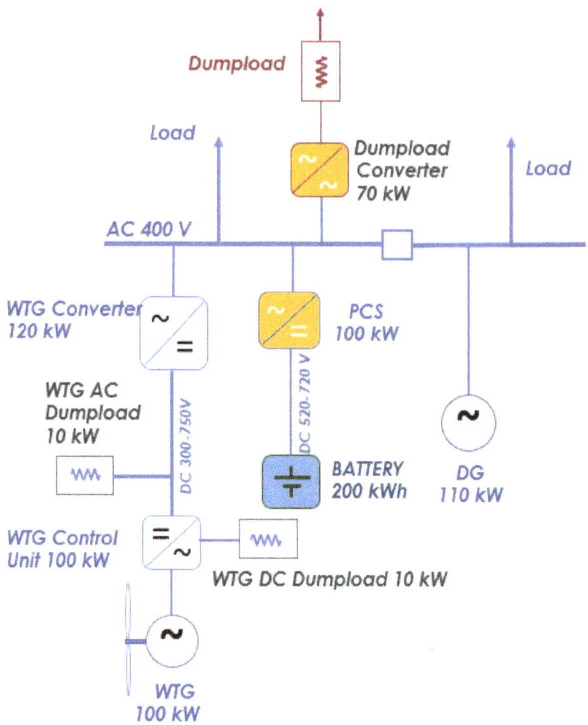

Figure 5. Structural scheme of the wind–diesel hybrid system.

The model used is compiled in Python (Python Software Foundation, Gemini Dr., Beaverton, OR, USA), on the core of a real wind–diesel power plant (WDPP) control system. The system has the following properties: a wind turbine of 100 kW, a full capacity converter, a diesel generator set of 110 kW, a battery energy storage system with a capacity of 200 kWh, a dump load with a capacity of 70 kW, and a real load graph (max 55 kW). The limitations of the model are as follows. Firstly, loading the initial data does not take into account the delay in their download. Secondly, the authors do not investigate the influence of the model operation speed on the signals of real facilities.

(1) Load following mode: In this mode, the diesel generator outputs electricity in accordance with the load (leading mode). The surplus electricity is first used to charge the battery and then to heat the water using the dump load. Disconnection of the diesel component is possible in the case of a fully charged battery and a prolonged excess of wind turbine output over the load. The diesel component is switched on when the battery voltage reaches the specified minimum. Thus, the battery works in deep cycles during periods of high winds. The dump load, together with the battery, contributes to the regulation of the network voltage to achieve its stable operation and acts as a buffer for load hesitation. Surplus energy is utilized in the form of useful heat for heating needs. Figure 6 shows hourly balances of power in supervisory control and data acquisition (SCADA)-based monitoring data over five days. From the figure, it is noticeable that for the majority of the time when the wind turbine is operating, the power balance significantly exceeds the load.
(2) Cycle charge with short-term forecasting mode: In this mode, the diesel generator works as an additional source of energy to cover power shortages. In the case of favorable wind turbine output forecasts, it is switched off. At the same time, the battery is used more efficiently and the size of the buffer capacity of the dump load

is reduced. The changes in dump load performance are visible when comparing Figures 6 and 7. From Figure 7, it can be seen that the diesel generator is repeatedly replaced by the battery discharge.

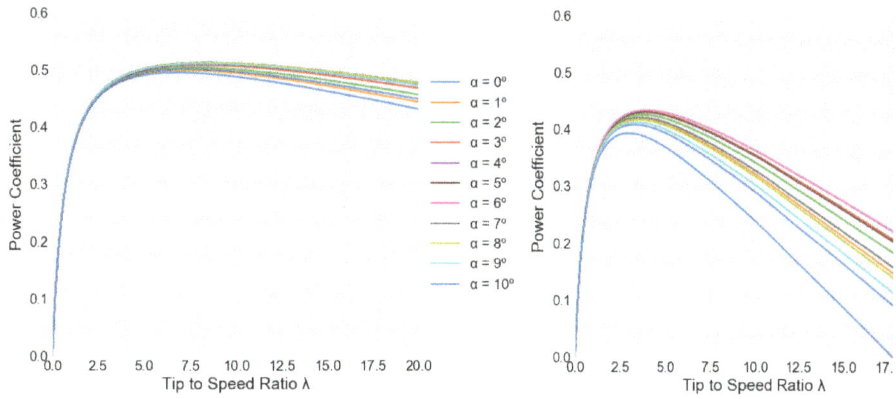

(a) (b)

Figure 6. Power balance under the load following mode (SCADA measurements).

Figure 7. Power balance under the cycle charge and short-term forecasting mode (simulation).

A more detailed comparison of the results is summarized in Table 3. The integration of the wind turbine and battery into the system during the analysis period facilitated fuel savings of 38%; however, to ensure the stable operation of the power system, a significant portion of the electricity generated by the wind turbine (48%) was distributed to the secondary regulatory load. With the wind turbine production forecasts and battery operation functioning in a cyclic mode, the share of wind turbine energy going to the secondary control load decreased to 38% and renewable energy penetration increased to 60%. At the same time, the number of battery cycles increased 2.2 times, up to two cycles of 80% charge/discharge per day (660 cycles per year). In their research employing similar system components (diesel generator, wind turbine, and battery), Elkadeem et al. [5] were able to reduce diesel fuel consumption by 85% (compared with a diesel-only system); however, their study had a significantly larger relative share of wind power capacity than the current study (more than two times the diesel generators' power). When compared with the load following mode, doubling the share of wind turbines seems to also roughly double the fuel savings. In relative terms, Li et al. [6] employed approximately similar diesel generator and wind power capacities but had significantly higher battery capacity (roughly three times higher). Their study reports a fuel saving of 74%. This leads to the conclusion that the proposed cycle charge with short-term forecasting mode can offer fuel-saving benefits by adding significantly more wind power or battery capacity without adding any actual new capacity.

Table 3. Operation statistics over five days.

Parameter	Diesel Only (Simulation)	WDDP Load Following Mode (Field)	WDPP Cycle Charge with Short-Term Forecasting Mode (Simulation)
DG energy output, kWh	3804	2423	1591.0
WTG energy output, kWh	0	3302	3302
Battery roundtrip output, kWh	0	450	937
Battery 80%, DOD cycles	0	4	9
DG start/stop cycles	1	3	13
DG specific fuel consumption, g/kWh	313	304	302
Fuel consumption, liters	1417	876	573
DG average load, %	31.5	35	31.5
Fuel savings, l	0	540 (38%)	844 (60%)
Dump load Energy consumption, kWh	0	1598	1247

3.2. *The Effect of Pitch and Tip-to-Speed Ratio Control*

To verify the chosen wind turbine icing modeling approach, the performance values of Table 2 were used to build power curves for clean and icing cases. The results of the modeling were then compared with the predictions from the Finnish Icing Atlas, which is based on the Finnish Wind Atlas [39] and ice aggregation modeling according to standard ISO 12394:2001. Since the figures in the Finnish Icing Atlas are reasonably sensitive to location, an area with a radius of 30 km was used to determine the maximum production loss values in each area for comparison with the exact location values presented in parentheses to illustrate local variations. The comparison reveals that the model presented in this work generally overestimates losses (Table 4). The magnitude of the predicted losses is, however, still similar to those achieved. This indicates that the proposed model can produce reasonable estimates for icing effects even though it is based on single airfoil data rather than data for full turbine blade shapes.

Table 4. Icing model comparison between the developed model and the Finnish Icing Atlas.

Location	Latitude	Longitude	Production Loss (%)	Icing Atlas Production Loss (%)
Olhava	65° N	25° E	7.1	5.9 (4.6)
Madetkoski	68° N	27° E	12.4	9.3 (4.0)
Jääräjoki	70° N	28° E	6.3	8.4 (1.5)

To implement the developed models in the energy system modeling tool, three curve fits were built based on the Madetkoski data from Finland. Figure 8 presents the power curves for a clean case, icing case, and pitch and tip-to-speed (optimized) control case and illustrates how the applied control approach can affect turbine performance. The total positive effect of optimized control on turbine power below the nominal operating point is between 2% to 5%. For wind turbines of medium and high capacity in Arctic zones, the optimized control system is advisable since the total cost of its installation is less than the total economic savings it can achieve. However, for wind turbines with a capacity of less than 300 kW, the installation of such a system must be confirmed by the relevant technical and economic analyses.

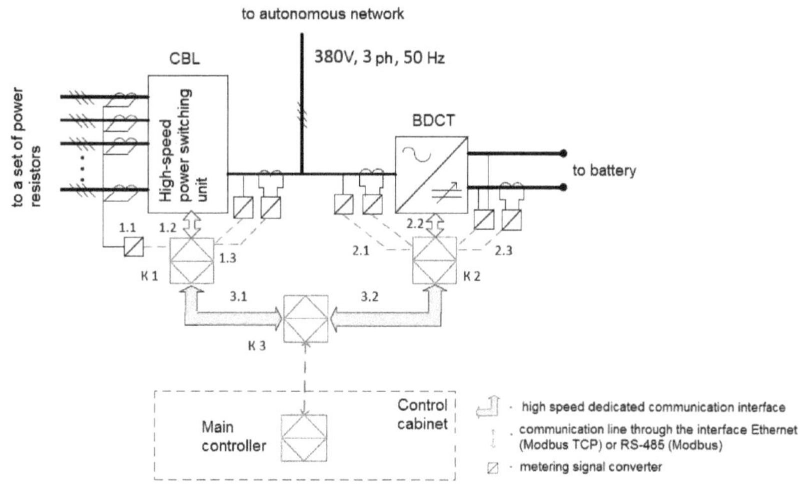

Figure 8. Wind turbine power curve under clean, icing, and optimized conditions.

4. Conclusions

1. The article presents the architecture of an advanced intelligent automatic control system for a wind–diesel hybrid system with high renewable energy penetration and describes the main modes of its operation.
2. The integration of the wind turbine and battery into the hybrid system enabled fuel savings of 38%, which were achieved by replacing the power generated by the diesel engine with wide turbine and battery power. In the load following mode, it was possible to disconnect the diesel generator when the battery was fully charged and wind turbine production was high. The dump load and the battery were used to regulate the network voltage. The excess electricity produced was used for heating.
3. With the addition of wind speed forecasting (LSTM model) and the cyclic charge mode, the share of wind turbine energy going to the secondary dump load decreased to 38%, and diesel fuel savings increased to 60%. Overall, the fuel savings correspond to the effects of significant additions of either wind turbine or battery capacities.
4. The net savings of using pitch and tip-to-speed ratio control exceed the cost of installing this system for medium- and high-capacity wind turbines. The use of the icing prediction unit in conjunction with weather forecasting and the turbine control system provides a more reliable operation of the wind turbine in harsh climatic conditions. It is estimated that these systems can reduce the operational expenditure (OPEX) by approximately 20%.

It is worth noting that the article does not examine the diagnostic unit. The current trend is the "complication" of data analytics towards deep machine learning and predictive diagnostics to prevent accidents through the application of accumulated experience and the analysis of large amounts of data.

The article considers the example of the onset of atmospheric icing; however, in the icing prediction block, it is necessary to calculate icing intensity using formula (1). In future research, these indicators will be studied in greater detail.

Author Contributions: The article is the result of the efforts of two working groups, one in St. Petersburg (SPbPU, Russia) and one in Lappeenranta (LUT University, Finland). Conceptualization, supervision, and project administration were overseen by V.E. and T.T.-S.; methodology, IACS-architecture, and visualization were overseen by R.D.; modeling and validation hybrid system modes were overseen by M.K.; modeling and validation of the pitch control system were overseen by A.J.L.

and A.G.; formal analysis, investigation, and formalization of results were overseen by I.B. All authors have read and agreed to the published version of the manuscript.

Funding: The research was carried out as part of the World-Class Research Center Program: Advanced Digital Technologies (contract No. 075-15-2020-934 dated 17.11.2020) and supported by the ENI CBC project KS1054, "Energy-efficient systems based on renewable energy for Arctic conditions".

Data Availability Statement: Nature data were gathered from the SCADA hybrid system project in Russian Arctic areas. Data for the icing calculations were collected from the Finnish Icing Atlas (Tammelin et al., 2011) and ice aggregation modeling according to standard ISO 12394:2001.

Conflicts of Interest: The authors declare no conflict of interest. The funders had no role in the design of the study, the collection, analysis, or interpretation of data, the writing of the manuscript, or the decision to publish the results.

References

1. Elistratov, V. Energy supply of autonomous territories based on renewable energy sources. In Proceedings of the 7th International Conference on Energy Efficiency and Agricultural Engineering, EE and AE 2020, Ruse, Bulgaria, 12–14 November 2020.
2. IEA. Projected Costs of Generating Electricity. 2020. Available online: https://www.iea.org/reports/projected-costs-of-generating-electricity-2020 (accessed on 30 April 2021).
3. Kazem, H.A.; Al-Badi, H.A.S.; Al Busaidi, A.S.; Chaichan, M.T. Optimum design and evaluation of hybrid solar/wind/diesel power system for Masirah Island. *Environ. Dev. Sustain.* **2017**, *19*, 1761–1778. [CrossRef]
4. Elistratov, V.; Kudryasheva, I.; Pilipets, P. Energy efficient solutions of power supply in north regions. *Appl. Mech. Mater.* **2015**, *725*, 559–568. [CrossRef]
5. Elkadeem, M.R.; Wang, S.; Sharshir, S.W.; Atia, E.G. Feasibility analysis and techno-economic design of grid-isolated hybrid renewable energy system for electrification of agriculture and irrigation area: A case study in Dongola, Sudan. *Energy Convers. Manag.* **2019**, *196*, 1453–1478. [CrossRef]
6. Li, C.; Zhou, D.; Wang, H.; Lu, Y.; Li, D. Techno-economic performance study of stand-alone wind/diesel/battery hybrid system with different battery technologies in the cold region of China. *Energy* **2020**, *192*, 116702. [CrossRef]
7. Elistratov, V.; Konischev, M.; Fedorov, M. Optimization of power supply of the circumpolar territories on the basis of renewable energy sources. In Proceedings of the International Conference on Industrial Engineering, Applications and Manufacturing, ICIEAM 2017, St. Petersburg, Russia, 16–19 May 2017.
8. Ansari, M.M.T.; Velusami, S. DMLHFLC (Dual mode linguistic hedge fuzzy logic controller) for an isolated wind-diesel hybrid power system with BES (battery energy storage) unit. *Energy* **2010**, *35*, 3827–3837. [CrossRef]
9. Vachirasricirikul, S.; Ngamroo, I.; Kaitwanidvilai, S. Coordinated SVC and AVR for robust voltage control in a hybrid wind-diesel system. *Energy Convers. Manag.* **2010**, *51*, 2383–2393. [CrossRef]
10. Elistratov, V.V.; Denisov, R.S. The optimization of hybrid systems' operating modes based on renewable energy. In Proceedings of the XVI-th International Conference on Electrical Machines, Drives and Power Systems ELMA 2019, Varna, Bulgaria, 6–8 June 2019.
11. Turkia, V.; Huttunen, S.; Thomas, W. *Method for Estimating Wind Turbine Production Losses Due to Icing*; VTT Technical Research Centre of Finland: Espo, Finland, 2013.
12. Wei, K.; Yang, Y.; Zuo, H.; Zhong, D. A review on ice detection technology and ice eliminationtechnology for wind turbine. *Wind Energy* **2020**, *23*, 433–457. [CrossRef]
13. Thompson, G.; Nygaard, B.E.; Makkonen, L.; Dierer, S. using the weather research and forecasting (wrf) model to predict ground/structural icing. In Proceedings of the IWAIS XIII, Andermatt, Switzerland, 8–11 September 2009.
14. Etemaddar, M.; Hansen, M.O.; Moan, T. Wind turbine aerodynamic response under atmospheric icing conditions. *Wind Energy* **2012**, *17*, 241–265. [CrossRef]
15. Bhatti, T.S.; Al-Ademi, A.A.F.; Bansal, N.K. Load-frequency control of isolated wind-diesel-microhydro hybrid power systems (WDMHPS). *Energy* **1997**, *22*, 461–470. [CrossRef]
16. Mi, Y.; Song, Y.; Fu, F.; Wang, S. The adaptive sliding mode reactive power control strategy for wind–diesel power system based on sliding mode observer. *IEEE Trans. Sustain. Energy* **2020**, *11*, 2241–2251. [CrossRef]
17. Mahto, T.; Mukherjee, V. A novel scaling factor based fuzzy logic controller for frequency control of an isolated hybrid power system. *Energy* **2017**, *130*, 339–350. [CrossRef]
18. Zavala, V.M.; Constantinescu, E.M.; Krause, T.; Anitescu, M. On-line economic optimization of energy systems using weather forecast information. *J. Process Control* **2009**, *19*, 1725–1736. [CrossRef]
19. Zhang, W.; Maleki, A.; Rosen, M.A.; Liu, J. Sizing a stand-alone solar-wind-hydrogen energy system using weatherforecasting and a hybrid search optimization algorithm. *Energy Convers. Manag.* **2010**, *180*, 609–621. [CrossRef]
20. Romero, A.; Carvalho, M.; Millar, D.L. Optimal design and control of wind-diesel hybrid energy systems for remote arctic mines. *J. Energy Resour. Technol.* **2016**, *138*, 062004. [CrossRef]

21. Elistratov, V.V.; Panfilov, A.A.; Konyschev, M.A.; Denisov, R.S. The application of adapted materials and technologies to create energy systems based on renewable energy sources under harsh climatic conditions. *Appl. Solar Energy* **2018**, *54*, 472–476. [CrossRef]
22. Elistratov, V.V.; Denisov, R.S. Energetic and ecological justification of RE-hybrid systems for vulnerable ecosystems. In *IOP Conference Series: Earth and Environmental Science*; IOP Publishing: Bristol, UK, 2021; Volume 689.
23. Elistratov, V.V.; Bogun, I.V.; Kasina, V.I. Optimization of wind-diesel power plants parameters and placement for power supply of russia's northern regions consumers. In Proceedings of the 16th Conference on Electrical Machines, Drives and Power Systems, ELMA 2019, Varna, Bulgaria, 6–8 June 2019.
24. Shukur, O.B.; Lee, M.H. Daily wind speed forecasting through hybrid KF-ANN model based on ARIMA. *Renew.Energy* **2015**, *76*, 637–647. [CrossRef]
25. Singh, S.N.; Mohapatra, A. Repeated wavelet transform based ARIMA model for very short-term wind speed forecasting. *Renew. Energy* **2019**, *136*, 758–768.
26. Carpinone, A.; Giorgio, M.; Langella, R.; Testa, A. Markov chain modeling for very-short-term wind power forecasting. *Electr. Power Syst. Res.* **2015**, *122*, 152–158. [CrossRef]
27. D'Amico, G.; Masala, G.; Petroni, F.; Sobolewski, R.A. Managing wind power generation via indexed semi-markov model and copula. *Energies* **2020**, *13*, 4246. [CrossRef]
28. Marugán, A.P.; Márquez, F.P.G.; Perez, J.M.P.; Ruiz-Hernández, D. A survey of artificial neural network in wind energy systems. *Appl. Energy* **2018**, *228*, 1822–1836. [CrossRef]
29. Santhosh, M.; Venkaiah, C.; Kumar, D.V. Ensemble empirical mode decomposition based adaptive wavelet neural network method for wind speed prediction. *Energy Convers. Manag.* **2018**, *168*, 482–493. [CrossRef]
30. Huang, C.-Y.; Liu, Y.-W.; Tzeng, W.-C.; Wang, P.-Y. Short term wind speed predictions by using the grey prediction model based forecast method. *IEEE Green Technol. Conf.* **2011**, 1–5. [CrossRef]
31. Gobiet, A.; Mitterer, C.; Jöbst, L.; Steinkogler, W.; Rieder, H.; Olefs, M.; Studeregger, A.; Monti, F.; Bellaire, S. Operational forecasting of wet snow avalanche activity: A case study for the eastern European Alps. In Proceedings of the International Snow Science Workshop, Breckenridge, CO, USA, 3–7 October 2016; Proceedings ISSW: Breckenridge, CO, USA, 2016; pp. 132–139.
32. Battisti, L. Wind turbines in cold climates: Icing impacts and mitigation systems. In *Green Energy and Technology*; Springer: New York, NY, USA, 2015.
33. Parent, O.; Ilinca, A. Anti-icing and de-icing techniques for wind turbines: Critical review. *Cold Reg. Sci. Technol.* **2011**, *65*, 88–96. [CrossRef]
34. Lugo, A.J. Pitch Control System for a Specific Arctic Wind Turbine. Master's Thesis, Lappeenranta University of Technology LUT, Lappeenranta, Finland. Available online: https://lutpub.lut.fi/bitstream/handle/10024/161310/AfonsoLugo_2020_MasterThesis_PitchControlSystem.pdf?isAllowed=y&sequence=1 (accessed on 30 April 2021).
35. Gantasala, S.; Tabatabaei, N.; Cervantes, M.; Aidanpää, J.-O. Numerical investigation of the aeroelastic behavior of a wind turbine with iced blades. *Energies* **2019**, *12*, 2422. [CrossRef]
36. Homola, M.C.; Virk, M.S.; Wallenius, T.; Nicklasson, P.J.; Sundsbø, P.A. Effect of atmospheric temperature and droplet size variation on ice accretion of wind turbine blades. *J. Wind Eng.* **2010**. [CrossRef]
37. Wilson, R.E.; Lissaman, P.B.; Walker, S.N. *Aerodynamic Performance of Wind Turbines*; Energy Research and Development Administration, Technical Information Center; Oregon State University: Corvallis, OR, USA, 1976.
38. Hudecz, A. Icing Problems of Wind Turbine Blades in Cold Climates. Ph.D. Thesis, Technical University of Denmark, Copenhagen, Denmark. Available online: https://orbit.dtu.dk/en/publications/icing-problems-of-wind-turbine-blades-in-cold-climates (accessed on 30 April 2021).
39. Tammelin, B.; Vihma, T.; Atlaskin, E.; Badger, J.; Fortelius, C.; Gregow, H.; Horttanainen, M.; Hyvönen, R.; Kilpinen, J.; Latikka, J.; et al. Production of the Finnish wind atlas. *Wind Energy* **2013**. [CrossRef]

energies

Article

Taking into Consideration the Inclusion of Wind Generation in Hybrid Microgrids: A Methodology and a Case Study

Luis Arribas [1,*], Natalia Bitenc [2] and Andreo Benech [3]

1 Renewable Energy Division, CIEMAT, 28040 Madrid, Spain
2 Engineering Department, Higher Technical School of Engineering and Industrial Design, Polytechnic University of Madrid (UPM), 28012 Madrid, Spain; natalia.bitenc@alumnos.upm.es
3 Autonomous Generation Department, National Administration of Power Plants and Electric Transmissions (UTE), Montevideo 2431, Uruguay; abenech@ute.com.uy
* Correspondence: lm.arribas@ciemat.es

Abstract: During the last decades, there has been great interest in the research community with respect to PV-Wind systems but figures show that, in practice, only PV-Diesel Power Systems (PVDPS) are being implemented. There are some barriers for the inclusion of wind generation in hybrid microgrids and some of them are economic barriers while others are technical barriers. This paper is focused on some of the identified technical barriers and presents a methodology to facilitate the inclusion of wind generation system in the design process in an affordable manner. An example of the application of this methodology and its results is shown through a case study. The case study is an existing PVDPS where there is an interest to incorporate wind generation in order to cope with a foreseen increase in the demand.

Keywords: design methodology; WDPS; microgrid; small wind turbine; wind data sources; HOMER Pro

Citation: Arribas, L.; Bitenc, N.; Benech, A. Taking into Consideration the Inclusion of Wind Generation in Hybrid Microgrids: A Methodology and a Case Study. *Energies* **2021**, *14*, 4082. https://doi.org/10.3390/en14144082

Academic Editors: Adrian Ilinca and Mohamed Benbouzid

Received: 29 April 2021
Accepted: 2 July 2021
Published: 6 July 2021

1. Introduction

Supplying remote isolated installations with diesel generators into a Diesel Power System (DPS) has traditionally been one of the most common solutions for all sizes of systems, but mainly for low (kWs range) and medium (up to MWs) power ones. With the development of renewable energies, their incorporation into the existing diesel grids started with wind energy due to its lower generation cost in comparison with solar PV, constituting the so called Wind Diesel Power Systems (WDPS) [1]. However, the twenty first century brought a strong reduction in the generation costs for solar PV, which has opened the door to its presence in existing DPS and constitutes the solar PV Diesel Power Systems (PVDPS). Actual trends transitioning to very high percentages of renewable energies (RE) at all levels of power systems induces the need of the Renewable Energy Diesel Power Systems (REDPS), where wind and solar PV technologies might seem to be the most upfront ones and where the presence of medium and long term (usually electrochemical) storage is common in order to reduce the use of fuel consumption.

In recent years, research has shown a growing interest in the use of hybrid wind photovoltaic (PV) systems. Over the past twenty-five years, hundreds of articles have addressed the topic of hybrid systems considering different configurations and final uses (a good representation of these papers can be found in the impressive literature review of photovoltaic-wind hybrid renewable system research by considering the most relevant 550 articles [2]) and, over the past decades, many reviews have made a comprehensive summary of various results obtained, which include, for example, the impressive review on more than 150 recent articles (including review and research articles) on sizing methodologies of hybrid renewable energy systems [3].

On the other hand, REDPS market (which somehow may be associated with Hybrid Systems and Microgrids ones) is well established and has grown during the last years.

However, the microgrids installed in the last five years have tended to incorporate PV and battery storage with diesel generators as backup (i.e., PVDPS or solar hybrid microgrids), coping with around 63% of existing microgrids [4]. On the other hand, small scale wind is much less common for microgrids although some combine with diesel or solar PV [5]. Thus, the common approach has become the following: "Solar hybrid microgrids are currently the most viable and reliable solution for off-grid areas with sufficient population and load density" [6].

These figures from the real world REDPS installations are in contrast with the previous figures from the research world, where there is a maintained interest in REDPS publications. This fact evidences a gap between research and the implementation of REDPS, showing that existing research in hybrid PV/wind microgrids consists primarily of feasibility studies, focusing solely on techno-economic aspects of PV/wind solutions [7], however, this is not sufficient for the implementation of such solutions in real applications. The reason for this might be relative to wind technology, however, nowadays wind technology is highly technically and economically developed at the large scale where large companies and consultancies are available for any design because the large size permits the necessary (not small) budget for the design process. The gap comes neither from the design methodology for PV/Wind solutions as there are plenty of examples of different design methodologies available in the literature, nor from the wind technology itself where implementation is widely spread at the large scale.

The following question arises: why is wind generation nearly missing in the micro-grid scale? The immediate answer would be that the dramatic cost reduction in the PV generation has taken it out of the playing field, but a pragmatic economic analysis shows that this is not always the case (that is, what the large number of related publications show). There must be other factors hindering the use of wind generation for this scale derived from the small size of the wind technology: Neither the technology nor the economic competitiveness are well achieved and the updated knowledge of this sector is not easy to maintain due to its constant impermanence. The impression coming from experience is that, in many cases, the small wind option is not even considered during the design process because of the lack and relative complexity of updated knowledge with respect to the technology ("implementation of the wind solution was discarded because the data analysis revealed a low energy production of the existing turbines ... ; the reason this is not working should be investigated" [8]), the wind resource ("Wind power is intermittent and its generation curve does not match the daytime load profile of communities" [4]), and its costs and main features ("A reduction in the cost of small-scale wind turbines at the same level as that seen with solar is not expected" [4]).

The main barriers identified in [7] for hybrid PV/wind microgrids in Kenya, which could easily be extended to most places in the world, are shown in Table 1, with some comments that will begin to show the scope of this work.

This publication is expected to provide some updated knowledge in order to facilitate the consideration of the inclusion of wind technology in the design of new or repowered REDPS of low power, which is not usually carried out by large consultancies and in which the budget for the design is usually limited by the size of the system.

Table 1. Categorization of the most commonly mentioned barriers for hybrid microgrids (source: Barriers and Explanations from [7]; comments are personal elaboration).

Barrier	Explanation	Comment
Wind resource quality	Wind resource is perceived to be insufficient.	See Section 2.1.3 (a), Sections 2.1.4, 3.1 and 3.3.
Technological knowledge	Knowledge of wind technology and capacity for development is low.	See Section 2.1.3 (b) and Section 3.3.

Table 1. *Cont.*

Barrier	Explanation	Comment
Governance and regulation	Lack of framework for implementation of wind power. Additional regulatory requirements for wind projects.	Yes, it is misunderstood. That is why benefits must be clearly understood.
Cost of wind projects	The cost of wind power is perceived to be higher than PV.	It may reduce the system LCOE.
Community acceptance	Understanding and acceptance of wind technology varies. Issues related to noise, bird incidents, and sickness as a result of turbines.	In general, these issues are much more critical in large scale wind: Information campaigns might help.
Complexity wind projects	Wind projects are seen as more complex. Necessary data are not available.	See Section 2.1.3 (a) and (c), Sections 2.1.4, 3.1 and 3.3.
Suitable technology	No suitable wind turbine exists. Low presence of local distributors or manufacturers.	See Section 2.2.1 (b).

2. Materials and Methods

A methodology to cope with some of the barriers that designers might face when considering small wind turbines as an option during the design process is proposed in this section.

A case study in the form of a feasibility study will then be presented as an example of the application of the proposed. The bulk of this example is not the techno-economic analysis of PV/wind solutions itself but the methods to cope with small wind technology issues within a more or less standard design procedure.

2.1. Methods: Methodology to Account for Small Wind Turbines Barriers during the Design Process

The reference design process will be briefly described (Section 2.1.2) once the configuration of the REDPS to be studied has been selected and justified (Section 2.1.1). Then, the main differences arising from the presence of wind generation in REDPS will be identified and the possible existing solutions for each identified factor will be analyzed (Section 2.1.3). Finally, the proposed methodology to account for small wind turbines barriers during the design process will be presented (Section 2.1.4).

2.1.1. Selection of the Configuration of REDPS to Be Designed

It is important to select the system configuration at the beginning because different configurations results in different design procedures and, thus, different considerations. The common feature of all the systems covered in this work is the presence of diesel generation. However, REDPS existing today can be classified from a technical point of view into diesel-dominated and inverter-dominated hybrid systems [9] depending on which component is in charge of maintaining grid stability.

- Diesel dominated hybrid systems.

Diesel engines have traditionally been one of the main options when electrifying rural and remote areas. However, the important drawbacks of this option (such as the rising cost of diesel fuel and carbon emissions concerns) when compared to renewable energy opens the door to the inclusion of such renewable energies in diesel dominated grids.

In these systems, the grid is formed by diesel generator(s). In fact, the common case is that there is an already existing grid supplied by the diesel engine(s) to which a renewable energy generating system is connected for retrofitting. Larger systems usually contain more and larger equipment that allows for an economy of scale and thus lowers power costs.

The system design is strongly related to the amount of energy that is expected from the renewable sources (system penetration), which will define the methods used to control the power system. System penetration can be defined either as Instantaneous penetration (renewable power output divided by the load power) or as Average penetration (renewable energy output divided by the total load energy over a given time period, typically a month or year). Instantaneous penetration relates to the power system complexity to maintain acceptable power quality. Average penetration relates to the steady state general system operation characteristics.

- Inverter dominated hybrid systems.

Most of the existing REDPS systems can be included in this group, which is characterized mainly by the inverters as the grid forming unit, and the use of long-term storage (in the past, usually lead acid batteries) as the main source to stabilize the grid. Fossil fuel generators are present but function as a back-up unit and usually with the options of supplying the load directly and bypassing the inverter if needed. From a configuration point of view, size is not such a great issue for this configuration but it is a fact that size is a very important issue for component availability, cost, and design.

The target configuration for this paper will be the inverter dominated REDPS, with one main diesel generator (there may be two for back up, but only one working at the same time) and with long term (several hours at least) electrochemical storage that is either AC or DC coupling.

2.1.2. Description of the Design Process

It may be worth remembering at this point that the methodology presented in this publication is not a design methodology, for which many other proposals have been published [3], but a methodology to cope with small wind turbine barriers during the design process. However, as this methodology is applied together with a design process, it will be briefly presented here.

From a design point of view, PV-Wind REDPS can be roughly considered as PVDPS where wind generation is added, as depicted in Figure 1. This approach has the advantage of allowing the use of all knowledge coming from existing solar PV standalone systems, which is quite extentsive.

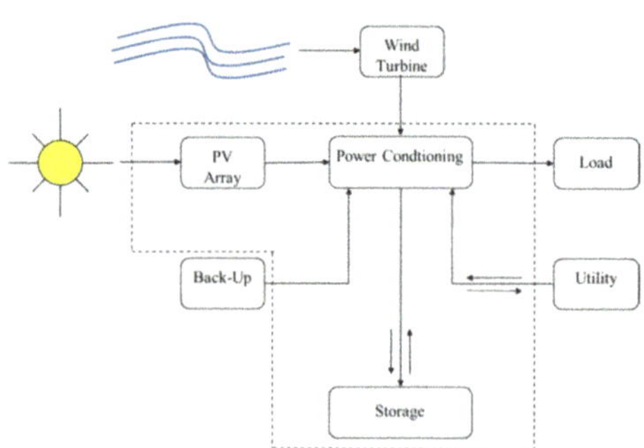

Figure 1. From a basic point of view, the design of Wind-PV-Diesel Power Systems may be seen as PVDPS where wind generation is added (adapted from [10]).

The research in REDPS might have started at the time when the IEA's Task 8 (1985–1994) "Study of Decentralized Applications for Wind Energy" aimed both to define cost-effective models and techniques suitable for obtaining wind and load data necessary for planning and specifying decentralized wind energy conversion installations and to apply and further develop models suitable for analyzing the performance of wind-diesel systems. The results of this collaborative research were summarized in the book "Wind-Diesel Systems", published by the Cambridge University Press [11]. At that time, the inclusion of solar PV generation was so expensive that it was not even considered. Since then, some implementation guidelines have already been proposed for both PV-Hybrid Systems [12] and Wind Hybrid systems [13,14], which both have many elements in common with PV-Wind Hybrid systems or even for small wind turbines (SWT) [15]. Some call for tenders for REDPS and offer their own design methodology [5].

From these guidelines, an adaptation has been made to update those particular issues that arise in PV-Wind DPS systems in three design stages: data collection, sizing study, and implementation Project.

2.1.3. Differences Arising from the Presence of Wind Generation in PV-Wind REDPS

It is a fact that the inclusion of wind generation results in different uncertainties in the design process of the system which needs to be taken into account. Some of the most important ones are addressed in the following paragraphs.

(a) *Characterization of Wind Resource*

Firstly, wind resource is much more variable (both temporarily and spatially) than solar resource and there is no geometrical method to predict it (in the case of solar resource, calculations of solar–Earth geometry provide a good estimation from latitude). This means that, on one hand, it is more difficult to evaluate and, on the other hand, small variations in the site may bring important variations in the wind resource.

○ *Wind Resource Assessment Methods*

There are a variety of methods to assess wind resource and they range from lower to higher costs. The methods of assessing the wind resource usually stem from using a general wind map and using that wind speed information to form the basis of a production estimate. Another approach is to use a commercial wind resource model to identify, more "precisely", the annual wind speed range. A third approach is to assess wind measurements from nearby projects, wind resource towers, airports, or other weather stations. Having equipment and tools that provide wind rose information are invaluable for the estimation of wind turbine production. Wind maps typically provide a basis for the wind speed, which is the other important factor in understanding production.

These three approaches do not take into account the impact of local micro-siting and the dramatic effect that it has on a small wind turbine. The only method where the wind resource is truly quantifiable and accounts for obstacle, terrain, wind direction, and blockages is to measure the wind at the exact location and exact hub height of the proposed small wind installation. Even though on-site wind measurements are the most reliable method of assessing the local wind resource, it is expensive and time-consuming and the cost is not justified for SWTs [16]. Historically the cost of wind measurement equipment and analysis has been prohibitive for small wind turbines. Recently new wind resource measurement approaches have been developed, including lower cost wind measurement equipment and towers and new drone technology [17].

Methods to cope with these cost restrictions are listed as follows: Using regional wind maps specifically for small wind turbines implementation, reanalysis data, or nearby Met office statistics; and choosing a site for the wind turbine as free from obstacles as possible.

○ *Using Reanalysis Data*

Given that the temporal variability of the wind, it important to know the time distribution over a period. Meteorological stations may be an option; however, they are not always close to the site or it is not possible to access their information. In that case, it is feasible to use reanalysis databases. Reanalysis data uses assimilation processes to combine observed (or measured) data obtained from satellites, ships, sensors, and weather stations with numerical models.

Since the observed data are unevenly distributed over the Earth, numerical meteorological models allow the estimation of the state of different layers of the atmosphere for a certain place and time period using a regular grid. With this approach, it is possible to generate a time series of gridded atmospheric parameters, such as air temperature, pressure, and wind at different altitudes; and surface parameters such as rainfall, soil moisture content, ocean-wave height, and sea-surface temperature. The three leading global data sets and its most recent bases are described below [18]:

- CFSR: This is the Climate Forecast System Reanalysis. Based on the Climate Forecast System, the NCEP global forecast model spans from 1979 to the present year. Most parameters are available every 6 h, whereas selected variables are available every hour. CFSRv2 provides wind speed and wind direction at 10 m above surface, with a horizontal resolution of 38 km.
- MERRA: MERRA is the Modern-Era Retrospective Analysis for Research and Applications, Version 2, based on the National Aeronautics and Space Administration (NASA) global data assimilation system (GEOS-5). MERRA-2 provides wind speed and direction at 50 m above surface. It holds one hourly values in a period from 1992 until present and a horizontal resolution of 50 km (0.5° latitude and 0.625° longitude). Hourly data for solar irradiation and wind speed can be accessed freely in the web service [19].
- ERA5: ERA5 is the fifth generation European Center for Medium-Range Prediction (ECMWF) atmospheric reanalysis of the global climate. It is a set of climate reanalysis data from 1979 to the present year developed by the Copernicus Climate Change Service (C3S) and processed by the ECMWF. ERA5 provides wind speed and wind direction at 10 m and 100 m above surface, with a special resolution of 31 km [20]. In the case of ERA5, data assimilation is performed every 12 h. However, it provides hourly estimations because the assimilation method takes into account the exact moment of the observations and the evolution of the model within the assimilation window.

Using reanalysis data also allows the easy performance of long term analysis on the viability of the inclusion of wind generation in hybrid diesel microgrid as data for decades.

○ *Small Wind Resource Assessment*

These databases have the advantage of having a high percentage of availability; however, they are not influenced by local effects of orography and roughness. Since roughness measures the decrease in wind speed due to friction with the surface and orography generating alterations in the wind flow, it is necessary to take them into account so that the energy assessment of the winds is as representative as possible of the site.

Using these global datasets to generate initial and boundary conditions for the simulations, some mesoscale and/or microscale models may produce higher resolution grids, such as WAsP ([21]). The most common mesoscale model is WRF [22], which is a weather prediction system used to generate meteorological forecasts or hindcasts. WRF downscales the global datasets and the results can be used to generate spatial wind maps and as a valuable source of long-term time series wind data. The WRF grid and output resolution is typically a few kilometers and ca. 3 km is often the preferred choice. It is possible to downscale the data from the global level to the mesoscale level and, further, to the microscale level, which is usually offered by commercial tools (as those from EMD or UL) with proprietary microscale models [23].

However, small wind turbines are often installed under wind conditions far from the conditions specified in standards and this is expected to result in large power curve uncertainties [17].

Complex terrain sites are typical small wind turbine sites and pose another challenge in accurately estimating wind turbine production. Underproduction was originally believed to be dominantly due to uncertified turbines and inconsistent turbine rating approaches; but as more turbines have become certified and wind turbine ratings are more globally consistent, underproduction is believed to be a strong function of the local micro wind conditions [16].

Over time, new modeling tools, site assessment technology, and study methodologies will evolve. Without streamlined customer-friendly approaches, the small wind turbine market will continue to be a smaller niche market. Better site assessment can be a step toward easing owner purchase decisions [16].

This paper is proposed for that aim.

(b) *Existing Technology*

There does not exist a unique definition for what a SWT is, in terms of size, but a more or less universal convention is that it refers to wind turbines smaller than 100kW. Within this range, some classification can be made according to Table 2.

Table 2. Classification of SWT (Source: CIEMAT).

Rated Power (kW)	Rotor Swept Area (m^2)	Sub-Category
Prated < 1 kW	A < 4.9 m^2	Pico wind
1 kW < Prated< 7 kW	A < 40 m^2	Micro wind
7 kW < Prated< 50 kW	A < 200 m^2	Mini wind
50 kW < Prated< 100 kW	A < 300 m^2	(No clear definition adopted yet)

The values that define the ranges for this classification have been chosen from the norms and legislation affecting SWTs. The value of 40 m^2 was the limit established in the first edition of the IEC-61400-2 standard and is the range intended at the present time for the integration of SWT into the built environment; the 200 m^2 limit was established in the second edition of the above mentioned IEC-61400-2 standard in 2006 and includes most SWT applications. Finally, the limit of 100 kW is defined in many countries as the maximum power that can be connected directly to the low voltage grid. The pico-wind range is commonly accepted as those SWTs smaller than 1 kW [24].

Despite being more uncertain, models and topographical background data are at a level of quality that makes the calculated wind resources valuable for SWT projects. Having determined the wind resources, the second source of uncertainty in energy yield calculations is the wind turbine type or more specifically the power curve of the wind turbine. Many countries today have standards for how wind turbine manufacturers should collect and process data to produce certified power curves; this improves the accuracy of the power curves that could otherwise be too "optimistic" [18].

Quality assurance has proven to be indispensable for establishing an enabling environment for a rapid uptake of renewable energy technologies. Quality assurance of standards are intended to ensure that products and services perform as expected and also includes the mechanisms to verify that such requirements are fulfilled, e.g., testing and certification [25]. This is of particular importance for SWT: A great effort has been conducted during the last two decades to increase the SWT quality and there are many reliable models in the market, but it is also possible to find many commercial models that have not been certified nor tested. A description of norms and standards affecting SWT can be found in [25] but, in general, an effort should be made to work with, at least, independently tested SWT. In [26], some guidelines can be found to assure quality for buying SWTs safely, which covers the manufacturer, the product reviews, the warranty, testing and certification and the installer.

Another important issue related to SWTs is the availability of wind turbines, which refers to both the available sizes and the available manufacturers. Before choosing a small wind turbine, it is advisable to be informed about sizes and maintenance support service in the area. This limits the number of available wind turbines for the design. An added difficulty is the highly changing characteristic of the SWT market (as an example, from the around 20 small wind turbines manufacturers in Spain in 2014, only eight are still active in this field in 2020 [27]). Information on available manufacturers and models should be updated frequently in order to be aware of the present situation. Both myWindTurbine [28] and HOMER Pro [29] include SWT databases, which are useful as a reference, but both of them are neither completely updated nor exhaustive. The reference, [30], may be a good starting point.

The last issue that will be highlighted here is the cost of the SWT. On one hand, there is a great variety in the cost of similar size commercial SWT: Cost should not be the only criterion to choose a SWT because, as it was mentioned before, not all of them will be of the same quality. On the other hand, the information on the cost of a particular SWT is not always easily accessible: Of course, the best method is always to have a particularized (the cost may be different according to the site) quote from the manufacturer/installer but some general references can be found in literature for a first approach [31–33].

(c) *Availability of Design Tools*

In general, when talking about available power systems design tools, there are different approaches depending on the following factors:

- Level of detail of the design, which is also related to the stage of the design of the project. When little detail is necessary, such as in the concept design phase, simple spreadsheet-type tools can be used, whereas dedicated tools are needed for a more detailed analysis, such as in the feasibility study stage. HOMER Pro is an international reference covering most of the levels of detail up to the feasibility study.
- Technologies involved: For example, whereas DPS are usually designed as a function of maximum and minimum power in the loads, REDPS usually takes into account the energy balance for their sizing, covering of course the power needs.
- Spatial resolution: Centralized REDPS can be designed with a unique profile for each RE resources and even a unique load profile (even if there are distributed loads, the generating system can be designed with a unique load and design the distribution lines), whereas distributed REDPS may require different profiles if distances are long enough. In this case, particular tools may be needed, including some Geographical Information System.
- Time resolution: different configurations of REDPS (see Section 2.1.2) require different time resolutions analysis. All of them require energy balance analysis, which are minimum monthly and preferably in a hourly time resolution, for the optimization design. Furthermore, diesel dominated architectures require dynamic analysis in the case of medium and high percentage RE systems. An example of used tools for a WDPS with high RE percentage is shown in Figure 2.

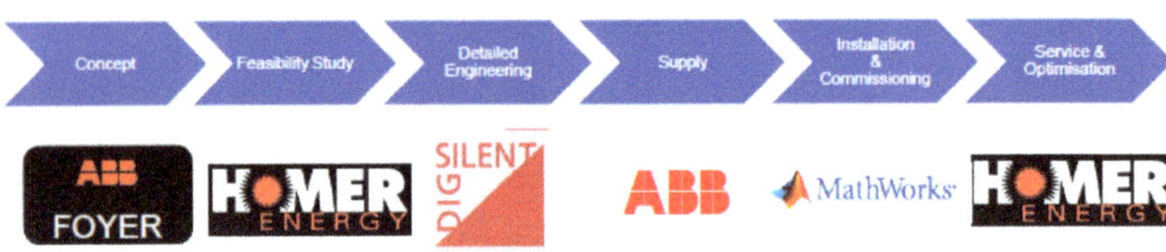

Figure 2. Example of tools used for a high RE penetration REDPS in the different design stages [34].

In the case of the selected configuration for this paper, dynamic analysis is not necessary for inverter dominated systems with one main diesel generator and with long term (several hours at least) electrochemical storage as the battery and the battery converter are able to maintain short term stability.

There is a great variety of sizing tools for PV Hybrid systems [35] but the number is limited when including wind generation in the system. Fortunately, even though the number of tools for PV-Wind REDPS design is limited, there are some high-quality available tools. According to a comparative study of 68 computer tools for the integration of renewable resource in various energy systems, HOMER was evaluated as one of the most applicable for optimization, feasibility, and sensitivity analysis of both off-grid and grid connected micro power systems and also pointed out as the most used and best known of all the software tools developed so far [8].

In particular, HOMER Pro has become a standard in the design of REDPS, as it merges most of the capabilities for a feasibility study in a single tool and it also includes wind technology (other well-known tools for PV systems design, such as PVSYST, does not); thus, it is the recommended tool for the configurations under study in this work. However, nowadays HOMER Pro software is no longer free (previously, there was a Legacy free version) and some amount must be paid depending on the desired use.

2.1.4. Methodology to Account for Small Wind Turbines Barriers during the Design Process

In the first place, the methodology described here deals only with some technical considerations of the design related to the consideration of wind technology; the development and management of a REDPS system is a relatively long and complicated process and involves other key aspects such as social, environmental, management, contractual, quality assurance, training, and some other aspects. These aspects obviously have to be taken into account through the project development, but are out of the scope of this work due to time and space limitations. Some useful references may be in [36,37].

As presented in Section 2.1.2, an adaptation has been made to cope with those particular issues that arises in REDPS systems during the consideration of wind technology, concluding with some suggestions that basically include the following stages.

(a) *Data Collection*

At this stage, the main inputs to the design process are gathered: topographic information, renewable resources (with special attention to wind resource, which is usually more difficult to evaluate), consumption characterization (crucial), and technical and economic information of the equipment. In projects of a certain size, it is a common practice to carry out field campaigns for both the wind resource and the consumption characterization [38].

In the case of performing a measuring campaign, in order to summarize the high quality requirements in wind measurement, international standards have been developed, such as the MEASNET guidelines. MEASNET is an international network of measurement institutes, which developed the guideline "Evaluation of site-specific wind conditions". This guideline describes the process of site assessment including data collection, evaluation, and interpretation. The MEASNET guideline refers to IEC 61400-12-1 and focuses on data quality, plausibility, and integrity [39].

Even for SWT there are some guidelines available for site assessment, such as [40], that covers most of the aspects that may apply when considering the inclusion of a SWT in a system. More detailed guidelines for site turbulence influence on the SWT estimated production can be found in dedicated guidelines, such as [15], for example.

However, one of the main limitations for taking into account wind generation in REDPS in the range that is being considered in this work is commonly the lack of reliable wind data necessary to evaluate its convenience. Nowadays, there are both global and local (usually at a national level) wind atlases. In this case, the no-data situation, which was not rare a few years ago, is almost extinguished. However, in order to perform a minimum performance evaluation of wind generation within a REDPS, it is not the case that any wind resource characterization will be valid: characterization needs to include spatial and temporal information in the most detailed possible manner. It should be remembered here that, of course, the best method to achieve this goal is an on-site wind measurement campaign, but it is precisely the case of not having this campaign that it is being covered at this point.

At least hourly average wind speed variation is needed to be able to assess the wind generation in a REDPS in order to evaluate its matching with load and other forms of generation profiles from an energy balance point of view. Other short-term phenomena (dynamics and/or transients) would require even higher time resolutions, but they are out of the scope of this particular application since there is long-term storage.

Global and local wind atlases may only bring information on the overall yearly wind regime [41] or even on the monthly average wind speed [42]. In the latter, some software (such as HOMER Pro) may generate hourly synthetic series from monthly average values of wind speed, so they might be a first approach. For the site assessment, wind direction estimation is also necessary. Global Wind Atlas [41] provides an overall wind rose.

The news on this topic is the availability of reanalysis data derived from satellite observations are described in Section 2.1.3 (a) in this paper, both globally and freely, and provides relatively sufficient time resolution (one hour) wind speed and wind direction data but with an insufficient spatial resolution of several square kilometers. However, there is a possibility to use these data as an input for downscaling (considering also the available information on roughness and DEM, Digital Elevation Models) by generating higher spatial resolutions down to several hundred meters. Although all these sources of information are freely available so that anyone could produce these results (these are good news for SWT), some commercial tools exist that allow performing it in an easier manner. As a reference, EMD in Europe and UL in the US offer different software solutions to obtain hourly wind resource estimation for any point (Windographer and WindNavigator, in the case of UL; WindPro, for EMD) with a reasonable spatial resolution: UL offers a Typical Year Time Series Short term data set using the AWST MASS model and scaled to 200 m resolution to represent a 365 day sample from a 15 year period [22], while with EMD's full windPRO modeling chain, it is possible to downscale the data from the global level to the mesoscale level and, further, to the microscale level (e.g., using the windPRO scaler options) to a 250 m resolution or even 100 m resolution with WAsP [21].

These two software groups also offer two particular applications that are especially suitable for the assessment of SWT in REDPS: myWindTurbine (EMD, [28]) was designed for the evaluation of the influence of obstacles in SWT production which, as it was mentioned in Section 2.1.3 (a), is a key issue for SWT; and HOMER Pro (recently acquired by UL, [29]), which is the reference for REDPS optimization, as it described in Section 2.1.3 (c). In its latest version of HOMER Pro 3.14.2 (10 August 2020), there is a link to WindNavigator ([43], UL software for wind resource estimation) that, while it is not yet available, opens the door to connect both applications.

One estimate of the cost to have a site modeled in the United States in 2016 was approximately $500. The model utilizes static wind maps, which is a gross approximation using annual average site wind speed and micro-site adjustments [17]. A characterization using myWindTurbine software costs around 60 EUR/site [28]. However, care should be taken with respect to the temporal needs in REDPS previously described, which may not be covered by these solutions.

(b) Sizing study

It is advisable to perform an initial screening of the components sizes at a simple spreadsheet level and at a conceptual design level: It is convenient to prepare the information for its use in the sizing tool (HOMER Pro is proposed, which in the last versions incorporates the Optimizer, and the automatic tool to search for the optimal configuration) and also it is convenient to have an estimate of the results. This spreadsheet may also be useful if a business model analysis is required, as HOMER Pro does not include it.

However, sizing tools require higher possibilities, such as the following: hourly basis simulation, which usually requires synthetic generation of data to compensate the possible lack of measured data; sensibility analysis, to cope with uncertainty of the inputs; more detailed (but still friendly) models of components, including all the necessary parameters to accomplish the study but at the lowest degree of complexity in order to make it useful; databases of commercial components, with the capability to create new ones.

At this level, both options of tools are available: commercial tools (either free or requiring a fee), such as HOMER PRO tool from Homer Energy by UL company, is an international reference; or self-designed proprietary tools, which take time and effort to develop but may be necessary if a specific analysis has to be made.

Real dynamic analysis is not usually necessary for these type of systems as they used to be inverter dominated microgrids based on a long-term battery using commercial power electronics solutions, which provide the necessary power quality and electrical stability to the system. However, it is convenient to implement this pseudo-dynamic analysis to assess the behavior of the battery through the simulation, mainly taking into account the influence of wind generation, as it was mentioned in Section 3.1. Whether this analysis is performed is up to the designer for establishing the stability of the system since the feasibility study does not cope with this issue.

The output of this stage in the design is a detailed behavior of the selected configuration in terms of stability and performance; the designer has to decide whether it is as expected or not (if not, then the designer would have to go back and make the necessary corrections on the configuration).

(c) *Implementation Project*

Once the configuration and equipment are settled, all of the components of the system are calculated and chosen through the "Sizing study" stage and it is time to prepare the necessary technical documentation for the deployment of the installation: schemes, plans, etc. This is purely an engineering stage; there is no dedicated software for this step, but programs of general use in engineering are commonly applied. However, some technology specific software may be necessary: for example, in order to design the PV generator (if it is present in the design), it is necessary to design the detailed configuration of the generator, which is not provided by HOMER Pro.

Even though this stage will not be covered in this paper, here are some hints in the case that some wind generation is present in the final design:

- Installation issues: foundation design; need and availability of a crane (which is related to the height and type of tower); available space. Manufacturer/installer should help with (or perform) this part.
- Electrical equipment: decide whether to use AC or DC coupling; power converter availability for the selected SWT and for the specific use; voltage and frequency (nominal value and range).
- Control issues: communication with the system control; dump load regulation (high temperatures), compatibility with Li-ion batteries (if present).
- Quality assurance: as it was described in Section 2.1.3 (b), all the related issues (guarantees and warranties, O and M, . . .) should be established.

It is worth talking, at this point, about the assessment of the installation. There is not enough experience on PV-Wind REDPS nowadays and it is important to make a provision for the monitoring system and its assessment. In reference [44], a new approach and a case study of a PV-wind hybrid system performance analysis is presented.

2.2. Materials: Case Study

The proposed methodology to account for the small wind turbines barriers during the design process has been applied on a real case appearing in Uruguay. The National utility (UTE) is in charge of an existing PVDPS deployed several years ago in order to reduce fuel consumption on the previously existing DPS. After several years of exploitation of the PVDPS, recently there was an interest from the utility in considering the inclusion of wind generation to increase both renewable energy penetration and load consumption. However, even though the UTE had been able to cope with the promotion and installation of the existing PV-Hybrid system, they faced the difficulties that including wind generation would bring, which demanded CIEMAT's experience in this field: this is how this real case study appeared.

In this section, the case study will be presented, showing the performance of the existing PVDPS through the analysis of data coming from the monitoring system as well. The application of the results derived from the use of the presented methodology will be shown in next Section 3.

2.2.1. Description of the Case Study: An Existing PVDPS in Cerros de Vera, Uruguay

Cerros de Vera is located in the northwest of Uruguay (Municipality of Salto), with approximately 70 electricity home services. Cerros de Vera is a small village that has been developed thanks to the support of MEVIR, an organization that works to eradicate unhealthy housing for rural workers by facilitating the construction and renovation of homes and access to community services. Since the village is not connected to the national grid, initially, the electrification was based on diesel generators. However, in February 2014, the town incorporated solar energy into its electrification system, becoming the first isolated rural village in Uruguay to be autonomously supplied with renewable energy.

The power system has an installed solar PV power capacity of 52.2 kWp (180 Suntech 290 Wp modules), two Kohler J88 diesel generators (64 kW each), and two battery banks (48 V, BAE made) with an overall storage capacity (C_{10}) of circa 300 kWh. The solar PV generator is divided into three groups, each one connected through a SMA Sunny Tripower inverter (17 kW each) to the 0.4 kV three-phase AC microgrid established by the two groups of three SMA Sunny Island 8.0 H power converters (3 × 8 kW each group). The average daily energy demand is around 440 kWh/day and the maximum demand is approximately 45 kW. In Figure 3, the general layout of the system is depicted. In Appendix A, some pictures of the system and its components are shown.

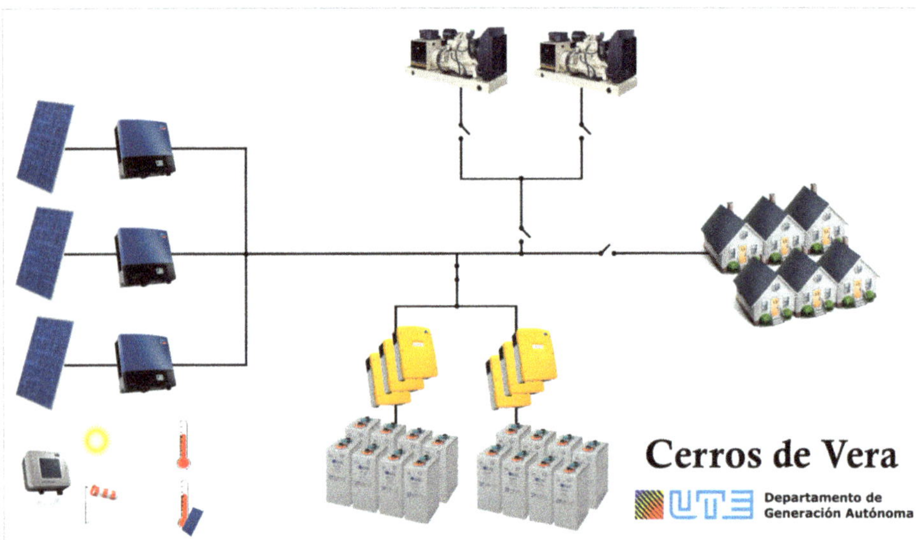

Figure 3. General layout of system under study, an existing PVDPS in Cerros de Vera, Uruguay.

2.2.2. Description of the Operation of the Existing System

The system is monitored through a weather station (connected to an SMA Sunny SensorBox) and the measurements are obtained from the power converters. Along with these measurements, the utility measures the energy consumption independently. From all these devices, data from 2018 and 2019 are available. After quality control of these data, a one year period from July 2018 to June 2019 was selected for evaluation. Sampling rate of data is 5 min for the raw data. A description of the system based on these data is shown now.

- Load Characterization.

From the measured data, the load patterns were assessed. In Figure 4, monthly (quite flat) and average hourly (peak in the nighttime) profiles derived from the measurements are shown, encouraging the hybridization of resources.

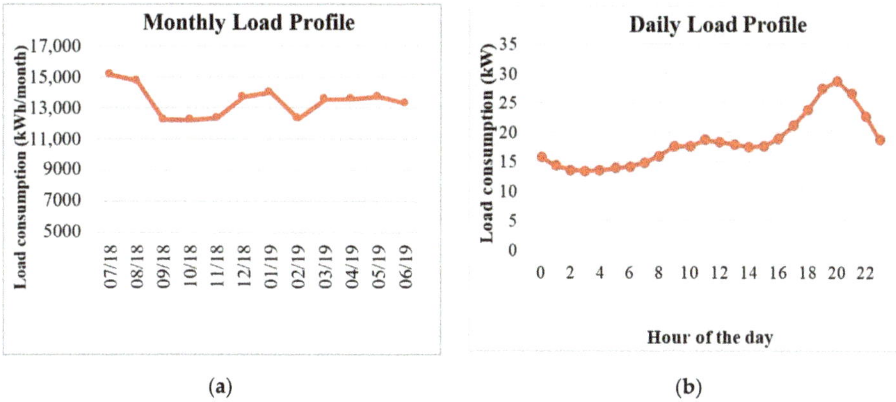

Figure 4. (**a**) Total monthly load consumption variation through a year; (**b**) average hourly load profile.

- Solar resource.

The weather station includes solar radiation measurements. From the measured data, the solar resource on-site has been derived, as shown in Figure 5. However, a validation of this resource has been made using available databases, as described later in the Data collection chapter.

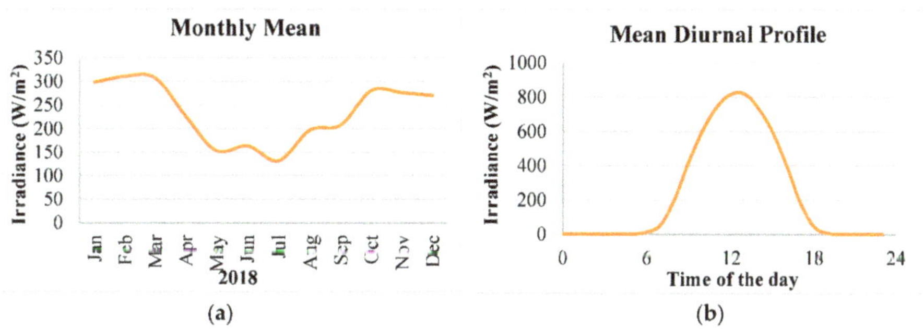

Figure 5. Average solar radiation variation (**a**) through a year; (**b**) hourly profile.

- Wind Resource.

The weather station also includes wind speed measurement, for solar PV generation purposes (not for wind generation purposes). The very low measured data in this case most probably will not be representative for wind resource assessment. In Figure 6, the measured results are shown and the validation of this resource is described later in the Data collection Section 3.1.

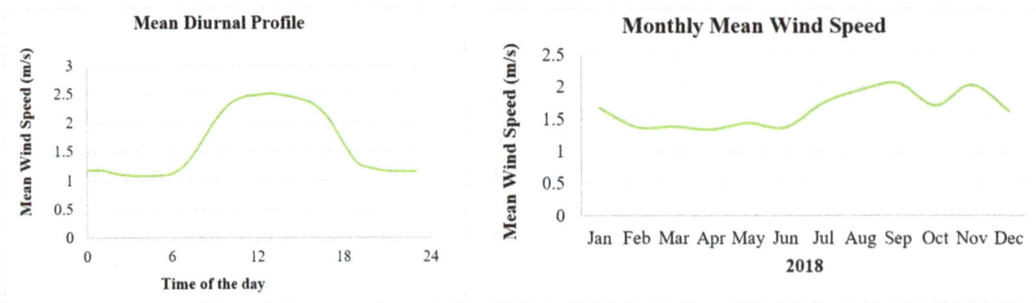

(**a**) (**b**)

Figure 6. Wind speed average variation: (**a**) hourly profile; (**b**) through a year.

- Photovoltaic system.

The measured data for the solar PV generator show an overall annual production of 64.4 MWh which, for a 52.2 kWp generator, represents 1233 equivalent hours. The measured performance ratio derived from measurements is 77.2%. This low value is explained because of the regulation of solar PV generation when the batteries are fully charged, as can be seen in the Figure 7.

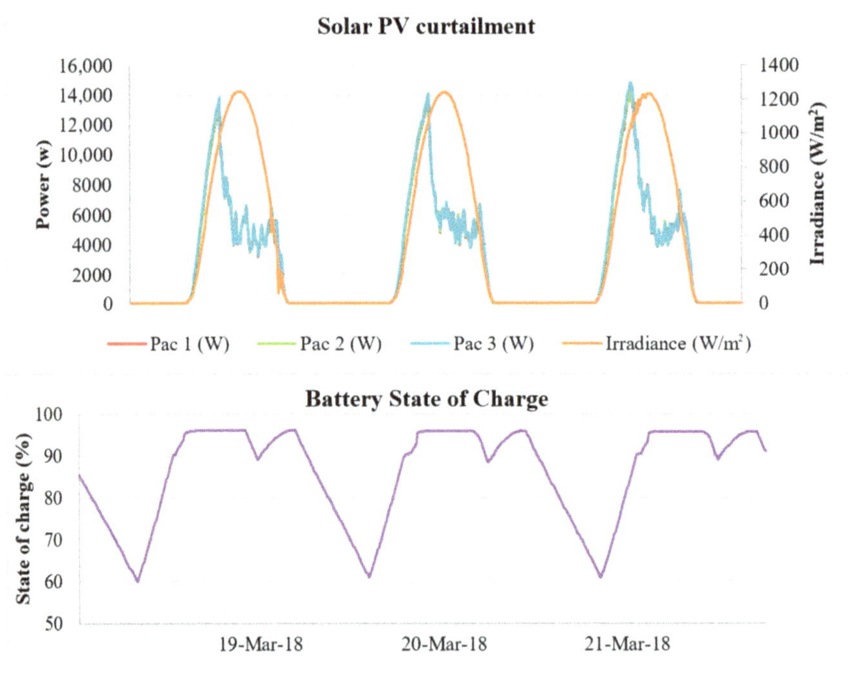

Figure 7. Solar PV curtailment when the battery is fully charged and are derived from measurements.

- Converters.

Both the overall average energy efficiency of the inverter and the rectifier have been calculated. The average values obtained are 92.33% and 89.71%, respectively (95.8% is the maximum efficiency in the specification sheets).

- Batteries.

The average efficiency has been calculated based on the energy received during charging and the energy delivered during discharge. As a result, an average efficiency of 77.1% has been obtained.

- Diesel generator.

Although there are two generators, only one of them works and the other one is used as a backup in case of breakdown. The system's current operating situation results in 40% of the energy generated resulting from solar energy, with diesel generators operating an average of 8 h a day. This implies a significant reduction in fuel consumption, but also in engine maintenance and extension of their useful life. Due to the facility´s design, the quality of service provided to the population was not affected.

3. Results: Application of the Proposed Methodology to the Case Study

In this section, the analysis of including wind generation in the existing system in Cerros de Vera will be covered using the proposed methodology. The application of the proposed methodology to the Cerros de Vera case study has been developed in three stages to analyze the feasibility of adding wind energy into the existing electrification system. The stages are: data collection, simulation of the operation of the current electrification system, and the generation and selection of future alternatives. They are sketched in the following diagram in Figure 8 and they will be described in the following subsections.

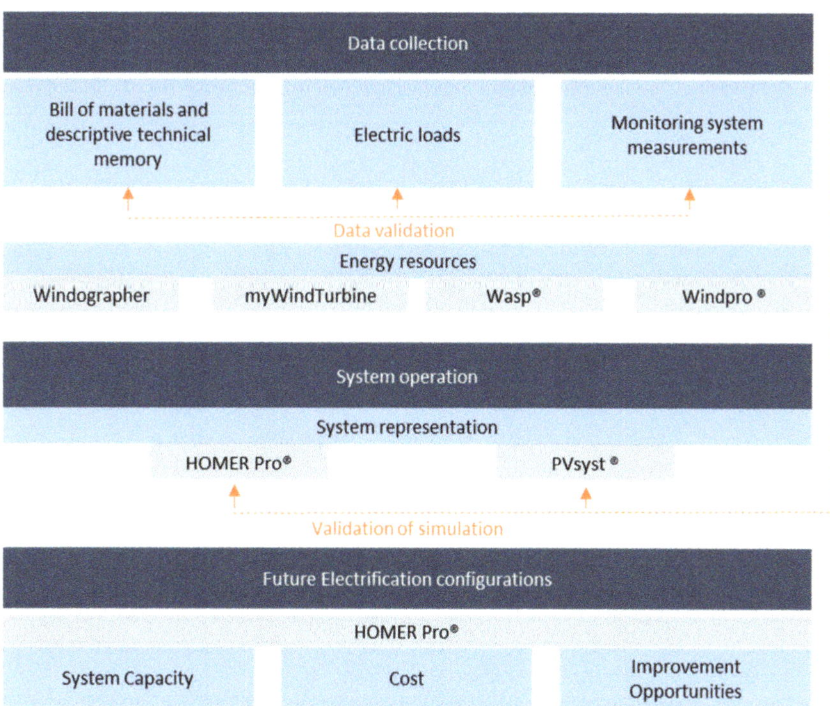

Figure 8. Flow chart for the proposed methodology applied to the case study.

3.1. Data Collection

The purpose of this stage is to collect objective information for the village necessary to perform the study. Since the case study is an existing project, sources for the collection of data are two-fold: most of the input data comes from equipment measurements developed in collaboration with utility UTE; on the other hand, public databases have been used to validate existing data, to gather data for optimization and, most important in this work, data for the inclusion of wind generation.

- Load demand data.

15-min records of active power consumption in the town have been obtained. Based on the data, the daily consumption curve, the annual profile, and the peak power have been calculated, as it was described in the previous chapter (see for example Figure 4).

- Solar resource.

The solar radiation is measured at the weather station installed. However, a validation with the following databases has been made: Meteonorm, PVgis, Tacuarenmbó, the nearest meteorological station, and TMY of Salto database. In the following graph in Figure 9, it can be seen how the data measured in the meteorological unit (Met Stat. 24°) are higher than the values of the public databases consulted. This discrepancy was most probably due to the different slope but, thanks to the comparison of the energy produced vs. the estimated with the measured irradiance, it has been possible to rule out the discrepancy and validate the data measured by the meteorological unit for its use in the analysis. These data were summarized in Figure 5.

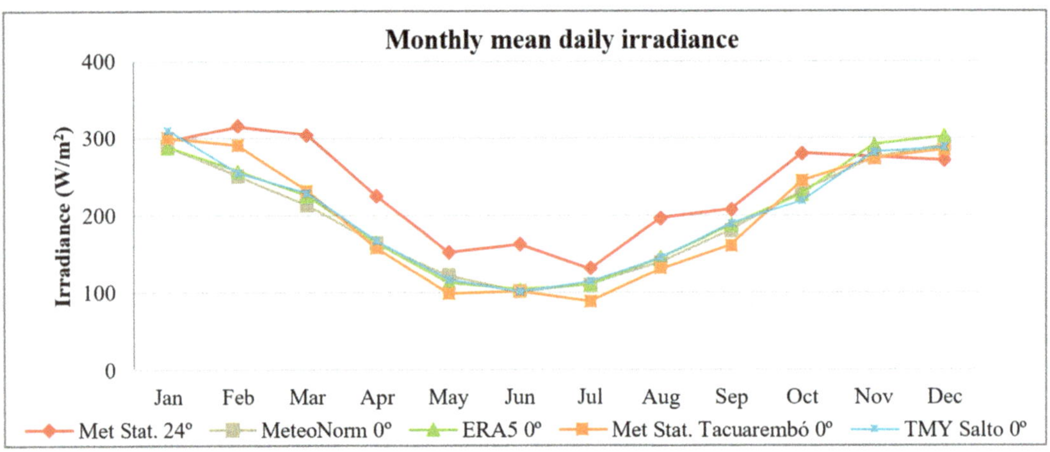

Figure 9. Comparison of measured data (Met. Stat. 24°) with databases (Meteonorm, ERA5, Tacuarembó Met. Stat., and TMY from Salto). Discrepancy comes from the different slope.

- Wind resource.

It was obtained from ERA5 database and downloaded with WindPro software. The selected node is located in 21S UTX 523,741 UTY 6,514,956. It is the closest to Cerro de Vera, with a distance of 12.8 km. The series used includes the period 2010–2020, with hourly data for wind speed and wind direction at 10 m, pressure, and temperature. Data obtained are not influenced by the local effects of orography and roughness.

The quality of the data obtained in the meteorological unit has been evaluated. Since the meteorological unit is designed to measure the variables that affect the photovoltaic installation, they have not been sufficient for characterizing the wind resource. These data have been supplemented with re-analysis databases. In order to evaluate the wind resource, two specialized programs in wind energy have been used; they include WAsP and WindPro.

Based on the ERA5 time series of wind at 10 m height, orography, and roughness, WAsP was used to perform a horizontal extrapolation that allows obtaining a resource grid. The grid has a resolution of 100 m (as the resolution of the used Digital Elevation Model is 90 m), with 48,816 points calculated at 10 m height. In this manner, it is possible to identify areas with better resources and therefore with greater energy potential to place the wind turbine. The resource grid has been imported into Windpro to continue the analysis from that software. Figure 10 shows the resulting resource grid on Windpro, the location of Cerros de Vera and the location of the ERA5 node.

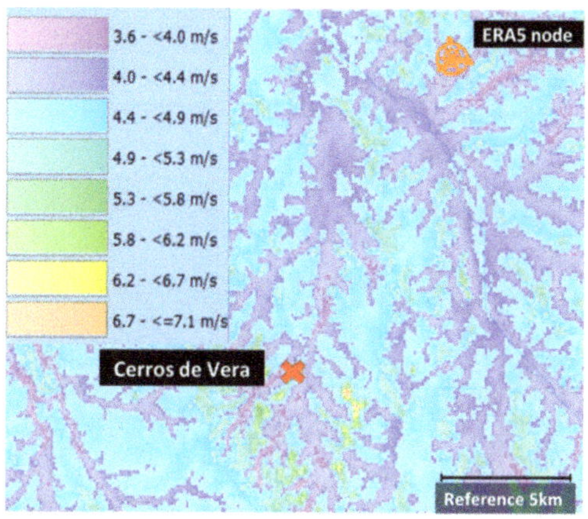

Figure 10. Resulting average wind speed at 10 m resource grid, with the location of Cerros de Vera on the bottom left and the location of the ERA5 node on the upper part.

- Temperature: obtained from ERA5 database along with wind data;
- Digital Elevation Model: obtained from SRTM 3 NASA database [45]. The downloaded map has an area of 40 × 40 km, with a 90 m spatial resolution. Additionally, a distance between contour lines of 5 m has been selected;
- Roughness: obtained from GlobalCover2009 −300 m database [46]. As for the orography, a 40 × 40 km map was downloaded. As well as the ERA5 wind data, the orography and roughness have been downloaded through Windpro. These outputs have been considered as the input for the creation of the resource grid;
- Technical aspects of the current electrification system.

The compilation of information on the existing system from the detail of the components to the costs (initial investment, operation and maintenance, fuel, etc.) and the operational experience as well. UTE has provided most of this information.

3.2. Simulation of the Operation of the Current Electrification System

This stage is particular of this case study, as it is an existing system. In order to simulate the future behavior of the system, the parameters for the simulation are adjusted so that they reflect the present performance in a more accurate manner. This stage seeks to determine the operation of the system, checking it against the simulation results system based on the information collected described in the previous chapter. Should there not be any existing system, only simulation results would be available, derived from estimated (not measured) information. At this point, the data from the monitoring system provided by UTE have been used.

Since Cerros de Vera has a photovoltaic installation, in order to determine the operating mode of the system, two software tools have been used: PVsyst® and HOMER Pro®. Since

PVsyst® is a software specifically designed for photovoltaic systems, greater precision is expected for the simulation of photovoltaic production and for this reason the results of PVsyst® will be used as input data for HOMER Pro®. The simulation of all system components will be carried out with HOMER Pro® using 1 h average data. Windographer® has also been used to process this information. The visual review of these data has allowed detecting anomalies in the operation of the different parts that will allow the improvement of the electrification system.

Below, after some first general considerations, the results obtained in the simulation for each of the components are described: photovoltaic system, converters, batteries, and diesel generator.

- General considerations.

The data provided by UTE for the load characterization have been compared with the monitoring system data.

The active power of the cluster and the photovoltaic power generated at the inverter output have been used to estimate the village´s consumption. As a result, a difference of 1.54% has been obtained. In Figure 11, the comparison for 4 September with a minimum variation of 0.7% is shown.

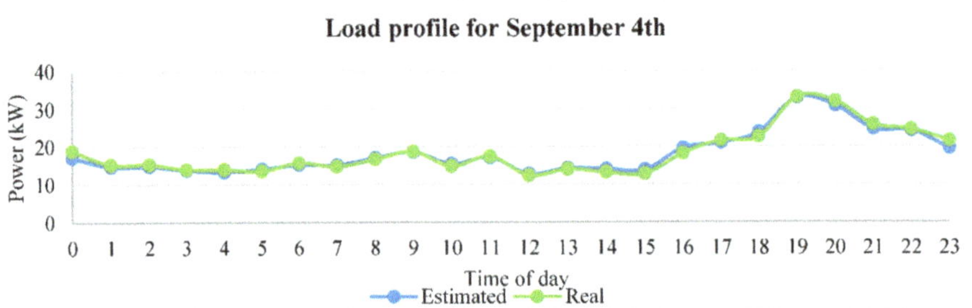

Figure 11. Comparison of load profiles according to the information provided by UTE meters (real) and by the power converters (estimated) for 4 September.

The consumption obtained from the data sent by UTE has been used as load input to HOMER Pro®. Additionally, the components used and their respective costs have been defined.

- Photovoltaic system.

Since PV production in excess is curtailed, a simulation with PVsyst® has been carried out to obtain photovoltaic production capacity without the influence of curtailment. For this purpose, the parameters of the PV generator generation indicated in the previous section have been adjusted.

In order to estimate the operation of the installation without curtailment distortions, the average performance ratio has been calculated for irradiation values lower than 800 W/m^2 and a module temperature between 22 and 28°. As a result, a PR of 80.4% has been obtained. This value has been used as a reference to estimate different parameters in the PVsyst simulation.

The simulation results show an energy production at the inverter output of 84,912 kWh/year, while the real energy produced was 64,403 kWh/year.

To verify if the parameterization carried out in PVSYST® corresponds with the monitoring system data, a comparison has been made showing a good correlation for the times without regulation, as shown in Figure 12.

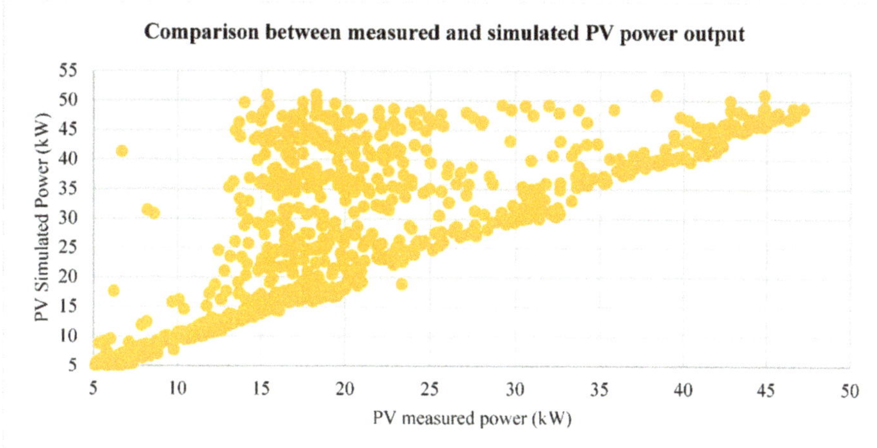

Figure 12. Comparison between the measured and simulated PV power output (points out of the linear correlation correspond to solar PV curtailment).

In order to compare the behavior of the rest of the components of the system, the generation corresponding to the real behavior, affected by the regulation, was kept as PV generation. However, in Section 3.3, corresponding to the optimization of the system, the production without curtailment has been analyzed in order to evaluate the possibilities for improvement.

The following comparisons are related to the period from October 2018 to December 2018.

- Converters.

As the Sunny Island 8.0H converter was not present in the HOMER Pro catalog, it was created. The average efficiencies for the inverter and rectifier were of 92.33% and 89.71%, respectively, based on the current behavior calculated from the data of the monitoring system.

- Batteries.

Even though the installation was defined with a depth of discharge of 70%, the value of 40% detected in the monitoring system for the validation period has been used. The battery efficiency has been considered 77.1% as calculated and the initial state of charge has been established at 97.87%.

When comparing the results of the simulation in HOMER Pro with the monitoring system data, only a difference of 1.96% in the charge (battery + rectifier) and 2.94% (battery + inverter) in the discharge has been obtained.

- Diesel Generators.

Regarding the operating mode, the "Cycle Charging" option has been selected where the generator works at full load to supply the village's consumption and, in case of electricity surpluses, the battery charge is supplied. According to measured data, the diesel generator is automatically forced to work from 19 to 22 h, independently of the state of charge of the battery to prevent the genset to automatically start during the night.

During the months from October to December, the energy delivered by the genset at the Cerros de Vera facility at the hours where data were available was 22,329 kWh, while the simulated power was 23,761 kWh, giving a difference of 6.41%.

A summary of the results expressed above can be seen in Table 3:

Table 3. Differences between the simulated and the measured results for the different components and for the validation period.

Component	Difference (%)
Photovoltaic system	4.8%
Converters + Batteries	1.96% charge/2.94% discharge
Diesel Generators	6.41%

Taking into account that the functioning strategy of the system is sometimes altered, according to measured data, these values are considered to be accurate enough, as it has been justified in this chapter.

3.3. *Analysis of Future Alternatives: Wind Generation*

An increase in the village loads and fuel consumption has been detected, triggering the following case study objectives: The analysis of the actual electrification system and the inclusion of wind energy in it. In this stage, the aim was to generate different electrification configurations considering what was evaluated in the previous points. Based on the technical and economic characteristics of the different alternatives obtained, this allowed the selection the optimal configuration. HOMER Pro has also been used at this stage. The design and analysis of the inclusion of wind generation to the current system in "Cerros de Vera" are detailed below.

- Wind Turbine Site.

Based on the analysis of the resource grid obtained from the ERA 5 wind data, which is detailed in Section 3.1, the wind turbine site has been selected. The criteria used was: wind resource and the proximity to the point of consumption and connections. Taking advantage of the fact that the photovoltaic installation has a security fence and enough space to locate the wind turbine inside, it has been decided to propose to place it within that area. The proposed site for the wind turbine is depicted in Figure 13.

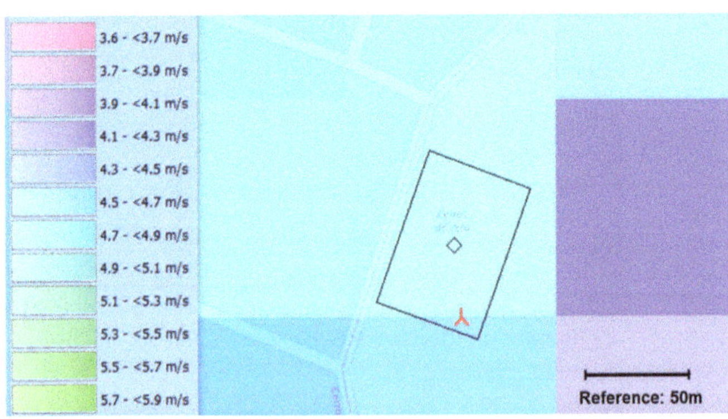

Figure 13. Resource grid at 10 m with the fence surrounding the solar PV area and the proposed location of the wind turbine.

- Influence of Obstacles.

In order to take into account the influence of the surrounding obstacles, the myWind-Turbine software has been used, identifying the main obstacles and using the software to evaluate the losses due to them. In Figure 14, the general view of the site facing East and the predominant wind direction are shown, along with the main obstacles considered for their assessment.

(a)

(b)

Figure 14. (a) General view to the predominant wind direction, East-Southeast; (b) Considered obstacles (in red).

The results of myWindTurbine analysis show a very small decrease in the annual energy production of 0.5% due to obstacles; the influence of obstacles will be neglected for this study.

- Wind Resource.

The wind resource assessment has been widely explained in the previous chapters. As a summary and neglecting the influence of obstacles in this case, according to the previous analysis the average speeds obtained in the site during 2019 for each month are detailed in Table 4 below and extrapolated at 24 m height using WAsP.

Table 4. Average monthly wind speed at the site at 24 m (extrapolated with WAsP from ERA5 10 m height wind speed data).

Month	Jan	Feb	Mar	Apr	May	Jun	Jul	Aug	Sep	Oct	Nov	Dec	Annual Avg
Wind Speed (m/s)	5.43	4.84	5.23	4.91	5.22	5.44	5.48	6.03	5.45	6.07	5.34	4.65	5.35

On the other hand, Figure 15 shows the wind rose obtained for the proposed site, showing East-Southeast as the clear predominant wind direction.

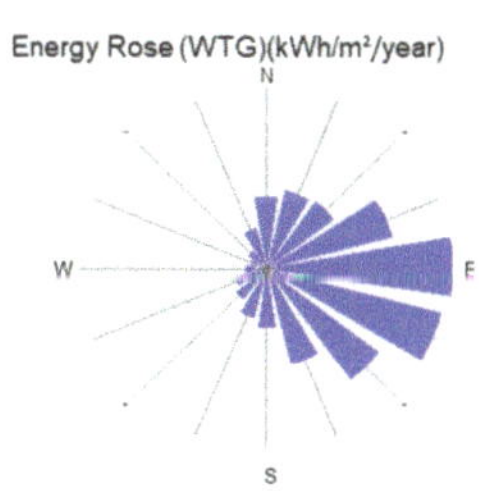

Figure 15. Wind rose on the proposed site showing East-Southeast as the predominant wind direction.

- Wind Turbine.

For the selection of the wind turbine, five wind turbines of different sizes have been evaluated. The sizes of the wind turbines selected correspond to 10 kW, 15 kW, 20 kW, 25 kW, and 30 kW. The established cost criteria are the following:

- Initial capital of 5000 EUR/kW for 10 kW turbines, 4000 EUR/kW for 30 kW turbines, and the costs are interpolated for the remaining sizes;
- Replacement cost: it has been defined as 85% of the initial capital value;
- Operation and Maintenance cost: a value of 45.5 EUR/kW has been assumed for 10 kW turbines and a value of 35.6 EUR/kW for 30 kW turbines. For the remaining sizes, it was obtained through interpolation.

In Table 5 the input variables defined to carry out the simulation are shown: the selected turbines, their nominal power, and hub height (Hh) together with the costs determined for the simulation.

Table 5. Wind turbines parameters.

Wind Turbine	Power (kW)	Hh (m)	Initial Capital (EUR)	Replacement Cost (EUR)	O and M (EUR/year)	System LCOE (EUR/kWh)
Bergey Excel 10	10	30	50,000	42,500	445.00	0.373
Gaia Wind 15	15	30	67,500	57,375	600.75	0.353
Eocycle E20	20	23	80,000	68,000	712.00	0.322
Ecocycle 25	25	23	87,500	74,375	823.25	0.320
PitchWind	30	30	90,000	76,500	801.00	0.332

Considering the overall performance of the wind turbines in this application at this site, the wind turbine finally selected from this analysis corresponds to the Eoclycle EO25 Class IIA [47] rated at 25 kW, using a 23 m hub height, with a LV connection to the system: It is not the highest and it is not the cheapest, but it results in the lowest overall system LCOE.

- Results of the Simulation.

The wind speed data together with the components (and sizes) of the current system and the selected wind turbine, reference values of the components, and the current consumption load of the village were considered as inputs to establish the configuration of the system in the simulation with HOMER PRO. As a remarkable fact, diesel fuel is currently being subsidized by the state of Uruguay; therefore, the cost for the diesel fuel is low (0.58 €/L).

Table 6 shows the result of the simulation of the current system operation (first row) and the optimal result considering a possible incorporation of wind energy in the current system (second row), showing the principal parameters for both cases.

Table 6. Results of the simulation for the alternative, including wind generation, compared to the existing installation.

Case	NPC (%)	LCOE (EUR/kWh)	Fuel Consumption (L/year)	Renewable Fraction (%)
Existing	100%	0.370	33,812	27.6
With wind	94.9%	0.320	16,717	64.2

When comparing both results, the positive impact of the inclusion of a wind generator in the system can be highlighted. The cost of energy fuel consumption is reduced and the renewable fraction exceeds 50%.

- A Sensibility Analysis.

A sensitivity analysis has been carried out to evaluate the impact that changes in the three main affecting parameters may have on the optimal configuration. The following variables have been considered: A variation considering a variation range in the annual average wind speed from 4.5 to 5.5 m/s (estimated value was 5.35 m/s); an increase of 5 and 15% in the village's consumption and a range in the price of fuel from 0.58 (present value) to 0.8 €/L. Figure 16 shows the impact of the variability of the wind speed and diesel fuel cost for the existing load consumption.

Figure 16 shows the relationship between the average wind speed at the site and the cost of fuel. The area in green represents the inclusion of wind power, while the black area represents the preservation of the current system. For wind speeds lower than 4.7 m/s and fuel costs lower than 0.7 €/L, the current system is optimal. Similar graphs can be derived for higher load consumptions, with a slight increase in the dark area as the energy consumption increases.

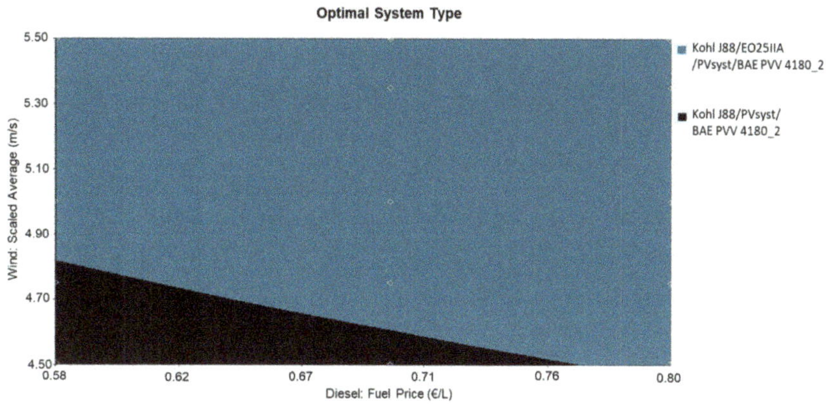

Figure 16. Sensitivity analysis result for diesel fuel cost (0.58–0.8 EUR/l, x axis) and average wind speed (4.5–5.5 m/s, y axis): The black area represents the existing configuration, whereas the blue are represents the configuration including wind generation.

In order to assist in the decision making, it is helpful to perform a long term analysis in order to have an estimation of the variation of wind speed compared to the period used during the analysis. This can be achieved easily using reanalysis data. In Figure 17, a histogram is presented, comparing the annual average wind speed values to the one used for the analysis (corresponding to 2018) for the last 42 years that is downloaded from ERA5 database. From this figure, at least two conclusions can be drawn: one, that the varying range is approximately ±5% in relation to 2018 (which might be helpful to move in Figure 16); second, that the year used as the reference for the simulation is in the average low area of the distribution and, therefore, it may be considered as relatively conservative (there are 14 years with lower average wind speed and 28 years with similar or higher wind speed).

As a concluding remark, the aim of this analysis is to raise a discussion on the different scenarios and to provide information for decision making.

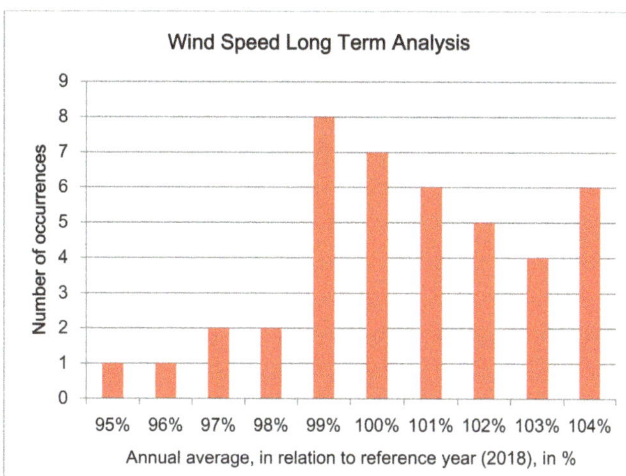

Figure 17. Long term analysis for annual average wind speed using ERA5 42 years' data.

4. Discussion

Having identified some technical barriers that hinders even the consideration of wind generation during the design process of REDPS, a technical methodology to cope with these barriers derived from the inclusion of small wind turbines in hybrid microgrids has been suggested. The target configuration has been established on inverter-based systems, with the possibility to include wind turbines of up to 100 kW (SWT) based on medium or long term (electrochemical) storage, with optional presence of solar PV generation which are very common nowadays.

With this target in mind, these differences have been focused on the wind resource characterization and on the description of existing technology, i.e., the small wind turbines. In relation to wind resource characterization, due to the need of expressing wind temporal variability (in a hour time frame, at least) and also due to the uncommon existing measuring campaign, a methodology to estimate the wind resource characterization from existing data bases and tools has been introduced. On the other hand, some description has been given for SWT technology in order to assure quality and expected performance of this component. In this manner, the system may be optimized (HOMER Pro software has been proposed for this) and necessary information is prepared in relation to the decision of moving forward with wind installation or rejection based on well-founded analysis and not based on myths relative to SWT.

5. Conclusions

This methodology proposed in this publication has been applied to a real case study, an existing PVDPS (installed in 2014) where the possibility of including wind generation is sought. Based on the performance of the existing system and on the characterization of the wind resource and wind technology, the design shows that there is a wide area in the search space of the main parameters (diesel fuel cost, average wind speed, and load consumption) where the installation of a 25 kW SWT seems recommendable.

Even though the aim of this paper is to present the methodology to permit wind technology consideration during the design process so that it can be implemented anywhere by anyone. In this particular case of application, it is expected that this study may be presented to the governing board of the utility in charge of the system so that they can evaluate the convenience of going forward with the project and to enter into the implementation of the project.

The proposed methodology used commercial software tools with a relatively high cost (mainly for the SWT field). The aim of the authors would be to achieve similar results with free tools but they are not available at the moment. One of the intentions of this paper is also to encourage researchers to keep working on such tools since it has been shown that it can be made.

Author Contributions: Conceptualization, L.A., N.B. and A.B.; methodology, L.A.; software, N.B. and L.A.; validation, N.B.; formal analysis, L.A. and N.B.; investigation, N.B.; resources, A.B.; data curation, N.B.; writing—original draft preparation, L.A. and N.B.; writing—review and editing, L.A., N.B. and A.B.; visualization, A.B.; supervision, L.A.; project administration, A.B. All authors have read and agreed to the published version of the manuscript.

Funding: This research received no external funding.

Institutional Review Board Statement: Not applicable.

Informed Consent Statement: Not applicable.

Data Availability Statement: The data presented in this study might be available on request from the UTE author. The data are not publicly available due to confidentiality issues.

Acknowledgments: The authors would like to thank UL for their support with information and to myWindTurbine.com (a dedicated tool for the planning of small domestic turbines, which was created as a joint effort between DTU Wind Energy and EMD International A/S.) for some free licenses.

Conflicts of Interest: The authors declare no conflict of interest. The support from the software companies had no role in the design of the study; in the collection, analyses, or the interpretation of data; in the writing of the manuscript or in the decision to publish the results.

Appendix A. Pictures from the System under Study

In this Appendix A, some pictures from the site and from the particular components of the existing system are shown.

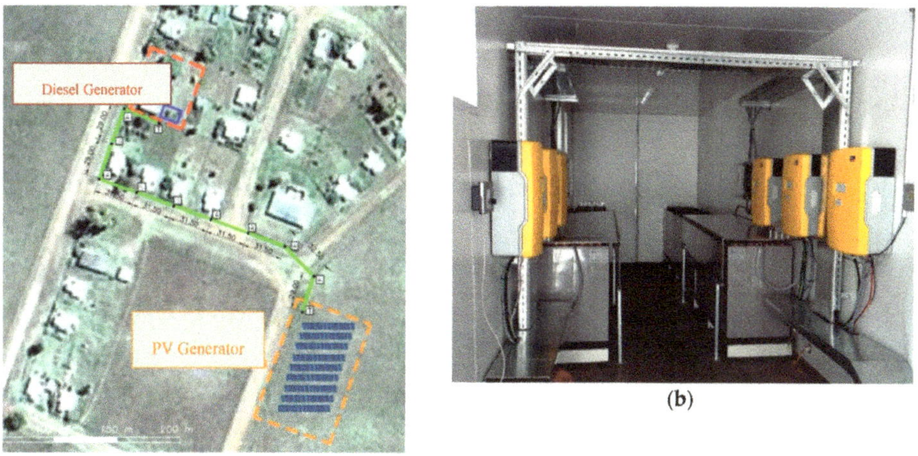

(a) (b)

Figure A1. (**a**) Aerial view of the site, with the indication of the existing solar PV array and the gensets building; (**b**) The two battery banks and the six Sunny Island power converters.

(a) (b)

Figure A2. (**a**) The inside of the gensets building with the two diesel generators; (**b**) A general view of the solar PV generator.

References

1. Baring-Gould, E.I.; Dabo, M. Technology, performance, and market report of wind-diesel applications for remote and island communities. In Proceedings of the European Wind Energy Conference, Marseille, France, 16–19 March 2009.
2. Mazzeo, D.; Matera, N.; De Luca, P.; Baglivo, C.; Congedo, P.M.; Oliveti, G. A literature review and statistical analysis of photovoltaic-wind hybrid renewable system research by considering the most relevant 550 articles: An upgradable matrix literature database. *J. Clean. Prod.* **2021**, 126070. [CrossRef]
3. Lian, J.; Zhang, Y.; Ma, C.; Yang, Y.; Chaima, E. A review on recent sizing methodologies of hybrid renewable energy systems. *Energy Convers. Manag.* **2019**, *199*, 112027. [CrossRef]

4. Bloomberg and SEforALL, 2020. State of the Global Mini-grids Market Report 2020. Available online: https://www.seforall.org/system/files/2020-06/MGP-2020-SEforALL.pdf (accessed on 3 July 2021).
5. Empresa de Pesquisa Energética (EPE). *Planejamento do Atendimento dos Sistemas Isolados*; Empresa de Pesquisa Energética: Brasília, Brazil, 2014. (In Portuguese)
6. ESMAP. *Mini Grids for Half a Billion People: Market Outlook and Handbook for Decision Makers*; World Bank: Washington, DC, USA, 2019.
7. Johannsen, R.M.; Østergaard, P.A.; Hanlin, R. Hybrid photovoltaic and wind mini-grids in Kenya: Techno-economic assessment and barriers to diffusion. *Energy Sustain. Dev.* **2020**, *54*, 111–126. [CrossRef]
8. Micangeli, A.; Del Citto, R.; Kiva, N.; Santori, G.; Gambino, V.; Kiplagat, J.; Viganò, D.; Fioriti, D.; Poli, D. Energy Production Analysis and Optimization of Mini-Grid in Remote Areas: The Case Study of Habaswein, Kenya. *Energies* **2017**, *10*, 2041. [CrossRef]
9. Lopes, L.; Mauch, K.; Katiraie, F.; Arribas. L. PV Hybrid Mini-Grids: Applicable Control Methods for Various Situations; Report IEA-PVPS T11-07:2012. 2012. Available online: https://iea-pvps.org/wp-content/uploads/2020/01/rep11_07.pdf (accessed on 27 September 2020).
10. International Electrotechnical Commission IEC 61724. *Photovoltaic System Performance Monitoring—Guidelines for Measurement, Data Exchange and Analysis*; Edition 1.0; International Electrotechnical Commission: Geneva, Switzerland, 1998.
11. Hunter, R.; Elliot, G. *Wind-Diesel Systems—A Guide to the Technology and its Implementation*; Cambridge University Press: Cambridge, UK, 1994. [CrossRef]
12. International Electrotechnical Commission IEC 62257-3. *Recommendations for Small Renewable Energy and Hybrid Systems for Rural Electrification—Part 3: Project Development and Management*; Draft Technical Specification Edition 1; International Electrotechnical Commission: Geneva, Switzerland, 2004.
13. Baring-Gould, I. Successful Project Development. In Proceedings of the Wind-Diesel Workshop, Ottawa, ON, Canada, 1–2 June 2009.
14. Clausen, N.E.; Bindner, H.; Frandsen, S.; Hansen, J.C.; Hansen, L.H.; Lundsager, P. *Isolated Systems with Wind Power An Implementation Guideline*; Risoe National Laboratory: Roskilde, Denmark, 2001.
15. Small Wind Guidebook. Available online: https://windexchange.energy.gov/small-wind-guidebook (accessed on 27 September 2020).
16. Bechmann, A.; Conti, D.; Davis, N.; Hansen, B.O.; Kelly, M.C.; Mortensen, N.G.; Nielsen, M.; Badger, J.; Pena Diaz, A. *MyWindTurbine—Energy Yield Calculations*; DTU Wind Energy: Roskilde, Denmark, 2016.
17. IEA Wind. *Task 27 Micro-Siting Small Wind Turbines for Highly Turbulent Sites*; Expert Group Report on Recommended Practices, International Energy Agency: Paris, France, 2018.
18. Brower, M.C.; Barton, M.S.; Lledó, L.; Dubois, J. Study of Wind Speed Variability Using Global Reanalysis Data. 2013. Available online: https://aws-dewi.ul.com/assets/A-Study-of-Wind-Speed-Variability-Using-Global-Reanalysis-Data1.pdf (accessed on 3 July 2021).
19. Renewables Ninja Web Tool. Available online: www.renewables.ninja (accessed on 3 July 2021).
20. ECMWF—Climate Reanalysis Web Page. Available online: https://www.ecmwf.int/en/research/climate-reanalysis (accessed on 30 September 2020).
21. WAsP Homepage. Available online: https://www.wasp.dk/ (accessed on 29 September 2020).
22. Weather Research and Forecasting Model. WRF Homepage. Available online: https://www.mmm.ucar.edu/weather-research-and-forecasting-model (accessed on 3 July 2021).
23. EMD International. EMD-WRF Global On-demand Mesoscale Services ERA5, ERA-Interim, MERRA2 and CFSR. Technical note. 2017. Available online: http://help.emd.dk/knowledgebase/content/TechNotes/TechNote_6_EMD-WRF_On-Demand.pdf (accessed on 3 July 2021).
24. Cruz, I.; Arribas, L. Small Wind Turbines. In *Wind Energy THE FACTS*; Routledge: London, UK, 2009.
25. IRENA. Quality Infrastructure for Renewable Energy Technologies Small Wind Turbines. 2015. Available online: https://irena.org/-/media/Files/IRENA/Agency/Publication/2015/IRENA_QI_2_SWTs_2015.pdf (accessed on 3 July 2021).
26. World Wind Energy Association. Quality Web Page. Available online: https://smallwind.wwindea.org/quality/ (accessed on 29 September 2020).
27. Arribas, L.; Barquero, C.; Cruz, I.; Avia, F. The Small Wind Turbines Market in Spain. Available online: http://www.ciemat.es/portal.do?TR=A&IDR=1&identificador=852 (accessed on 3 July 2021). (In Spanish)
28. myWindTurbine Homepage. Available online: https://www.mywindturbine.com/ (accessed on 29 September 2020).
29. HOMER Energy. HOMER (Hybrid Optimization of Multiple Energy Resources) Homepage. Available online: https://www.homerenergy.com (accessed on 29 September 2020).
30. Folkecenter; 2016 Small Wind Catalogue. Available online: http://folkecenter.eu/pages/Small-wind-turbine-catalogue.html (accessed on 3 July 2021).
31. Orrell, A.C.; Rhoads-Weaver, H.E.; Flowers, L.T.; Jenkins, J.O.; Gagne, M.N.; Sahl, K.M.; Pro, B.H.; Baranowski, R.E. 2012 Market Report on U.S. Wind Technologies in Distributed Applications. Available online: https://www.pnnl.gov/main/publications/external/technical_reports/PNNL-22537.pdf (accessed on 30 September 2020).

32. Woofenden, I. 2014 Wind Turbine Buyer's Guide. In *Home Power Magazine 161*; Oregon, USA, 2014. Available online: https://www.scribd.com/document/481612016/hp-161-pdf (accessed on 30 September 2020).
33. Sigrin, B.; Gleason, M.; Preus, R.; Baring-Gould, I.; Margolis, R. *Distributed Generation Market Demand Model (dGen): Documentation*; NREL/TP–6A20-65231; National Renewable Energy Laboratory: Golden, CO, USA, 2016.
34. Zimmermann, J. Case study for CORAL BAY Energy modelling for a WIND DIESEL and stabilization hybrid system. In Proceedings of the Microgrid Deployment Workshop Fall, Barcelona, Spain, 20 September 2014.
35. Arribas, L.; Bopp, G.; Lippkau, A.; Mauch, K. World-Wide Overview of Design and Simulation Tools for Hybrid PV Systems. Report IEA-PVPS T11-01:2011. 2011. Available online: https://iea-pvps.org/wp-content/uploads/2020/01/rep11_01.pdf (accessed on 27 September 2020).
36. The World Bank. *Designing Sustainable Off-Grid Rural Electrification Projects: Principles and Practices*; The World Bank: Washington, DC, USA, 2008.
37. Jacquin, P.; Ortiz, B.; Vallvé, X. Social, Economic and Organizational Framework for Sustainable Operation of PV Hybrid Systems within Mini-Grids. IEA PVPS Task 11, Report IEA-PVPS T11-05:2011. 2011. Available online: https://iea-pvps.org/wp-content/uploads/2020/01/rep11_05.pdf (accessed on 27 September 2020).
38. UNDP. Wind Diesel Generation Systems for 10 Islands in the Chiloe Archipelago. UNDP International call for Proposals 023/2007. 2007. Available online: http://procurement-notices.undp.org/view_notice.cfm?notice_id=1193 (accessed on 3 July 2021). (In Spanish)
39. MEASNET. Evaluation of Site-Specific Wind Conditions. MEASNET Procedure. Version 2. April 2016. Available online: https://www.measnet.com/wp-content/uploads/2016/05/Measnet_SiteAssessment_V2.0.pdf (accessed on 3 July 2021).
40. Olsen, T.; Preus, R. Small Wind Site Assessment Guidelines. Technical Report NREL/ TP-5000-63696. September 2015. Available online: https://www.nrel.gov/docs/fy15osti/63696.pdf (accessed on 3 July 2021).
41. Global Wind Atlas Homepage. Available online: https://globalwindatlas.info/ (accessed on 29 September 2020).
42. POWER Project Homepage. Available online: https://power.larc.nasa.gov/ (accessed on 29 September 2020).
43. Wind Navigator Homepage. Available online: https://aws-dewi.ul.com/software/windographer/windnavigator/ (accessed on 29 September 2020).
44. Arribas, L.; Cano, L.; Cruz, I.; Mata, M.; Llobet, E. PV–wind hybrid system performance: A new approach and a case study. *Renew. Energy* **2010**, *35*, 128–137. [CrossRef]
45. Shuttle Radar Topography Mission. WindPro Knowledgebase. Available online: https://help.emd.dk/mediawiki/index.php?title=Shuttle_Radar_Topography_Mission (accessed on 23 June 2021).
46. Global Land Cover Characteristics. WindPro Knowledgebase. Available online: https://help.emd.dk/mediawiki/index.php?title=Global_Land_Cover_Characteristics (accessed on 23 June 2021).
47. Eoclycle. EO25 Class IIA Wind Turbine. Specification Sheet. 2020. Available online: http://energieclub.weebly.com/uploads/1/1/6/7/11670442/eocycle-data-sheet-eo2512-may13.pdf (accessed on 30 September 2020).

energies

MDPI

Article

Long-Range Integrated Development Analysis: The Cuban Isla de la Juventud Study Case

Ernesto Alberto Alvarez [1], Mika Korkeakoski [2,*], Ariel Santos Fuentefría [1], Miriam Lourdes Filgueiras Sainz de Rozas [1], Ramsés Arcila Padura [1] and Jyrki Luukkanen [2]

1. Faculty of Electrical Engineering, Electroenergetic Research and Testing Center (CIPEL), Technological University of Havana, CUJAE, 11500 Havana, Cuba; ealvertoa@electrica.cujae.edu.cu (E.A.A.); asfuentefria@electrica.cujae.edu.cu (A.S.F.); miriaml@electrica.cujae.edu.cu (M.L.F.S.d.R.); rapadura@cujae.edu.cu (R.A.P.)
2. Finland Futures Research Centre Turku School of Economics, University of Turku, UTU, 20014 Turku, Finland; jyrki.luukkanen@utu.fi

* Correspondence: mika.korkeakoski@utu.fi

Abstract: The use of renewable energy sources (RES) has increased exponentially worldwide, as an alternative to the indiscriminate use of fossil fuels and to mitigate their effects on the environment. Cuba is not lagging behind in this development since the government's plan until 2030 includes the contribution of renewable sources as a fundamental component in the national energy mix. This paper models possible scenarios based on 2019 statistics for achieving a 25% and 100% penetration of renewable sources by 2030 in the Isla de la Juventud's (an island south of the main island of Cuba) electrical power system (EPS). This modeling is carried out utilizing and open source Excel-based accounting framework Long-range Integrated Development Analysis (LINDA). For this purpose, international and national trends in the use and development of renewable energy sources and the influence of the characteristics of each renewable source (wind, solar, biodiesel, battery storage) were analyzed. The analysis of Isla de la Juventud's electrical power system was based on the characteristics of its energy mix, the possibilities of renewable energy penetration and the current and future energy demand by sector. Based on the analysis, two probable scenarios were modeled with LINDA model: a 25% renewable energy-based scenario (RENES) and a 100% renewables-based scenario (MAXRES). Results from RENES and MAXRES scenarios show high penetration of renewable energy sources in electricity generation is theoretically possible with the abundance of renewable energy resources, and thus it is possible for Cuba to move towards 100% renewable energy mix. However, the choices regarding the best fit energy mix need to be carefully analyzed in order to design a least cost system that answers the needs of the future demand.

Keywords: Isla de la Juventud; electrical power system; renewable energy; long-term planning; LINDA model

Citation: Alberto Alvarez, E.; Korkeakoski, M.; Santos Fuentefría, A.; Lourdes Filgueiras Sainz de Rozas, M.; Arcila Padura, R.; Luukkanen, J. Long-Range Integrated Development Analysis: The Cuban Isla de la Juventud Study Case. *Energies* **2021**, *14*, 2865. https://doi.org/10.3390/en14102865

Academic Editor: Rafael Sebastián Fernández

Received: 19 April 2021
Accepted: 6 May 2021
Published: 15 May 2021

1. Introduction

Climate change is widely considered as the greatest challenge facing humanity and growing emissions, especially from the energy sector, are the main drivers of global climate change [1]. Although the use of renewable energy sources globally has grown dramatically, the continued reliance on fossil fuels has resulted in emission increase of nearly 2% to 33.1 Gtons of CO_2 in 2018 [1].

In view of these facts, specialists and governments around the world have become increasingly aware of the importance of addressing climate change through the use of renewable energy sources (RES) and efficient energy management. Renewable energies are also becoming increasingly competitive as they are clean and inexhaustible sources with marked differences from fossil fuels, mainly because of their diversity, abundance, potential for use anywhere on the planet and, above all, they do not produce greenhouse gases. The global weighted-average levelized cost of electricity (LCOE) of utility-scale solar photovoltaics (PV) decreased by 82% between 2010 and 2019 and onshore wind by 39% according International Renewable Energy Agency (IRENA). The same trend is likely to continue in the short-term and out to 2030 [2]. Just in 2018 the share of electricity produced from renewables grew by over 7% [1]. The increased use of renewable energy sources is facilitating new economic opportunities and access to energy for millions of people who still live without electricity services. According to the United Nations, in 2018 11% of the world's population has no access to electricity [3]. The amount of population without access continued to decrease in 2019 from 860 million in 2018 to 770 million in 2019 [3,4]. Achieving a universal access to affordable, reliable, sustainable and modern energy for all is one of the Sustainable Development Goals (SDGs) set by the United Nations. [5]

In line with the efforts made worldwide, Cuba adopted a new program in 2011 to modernize and strengthen the electricity sector, promoting the use of different renewable energy sources, mainly biogas, wind, hydro, biomass and solar energy [6]. This led, in 2017, to a policy for the "development of renewable energy sources and energy efficiency". The main objective of this policy is to increase generation by renewable sources of energy to 24% of the primary energy sources by 2030 through:

- Transforming the energy mix with a greater share of renewable sources and other national energy resources.
- Creating a reliable, diversified, environmentally sustainable and modern energy supply, at competitive prices and substantially increasing the share of renewable energy sources (essentially biomass, wind and solar) in the national energy mix [7].

At the end of 2018 the Cuban energy production was highly dependent on fossil fuels, with around 95.5% of production coming from fossil fuels and only 4.5% from renewable energy sources [8]. The national electrical power system has been structured through a combination of condensing power plants and combined heat and power (CHP) baseload, diesel and fuel oil decentralized power generation, bioenergy from sugarcane bagasse and small amounts of power from biogas, hydro, solar and wind sources. In total, in 2013 renewables accounted only for 4.3% of the total electricity production of the country [8].

Cuba has a vast renewable energy potential to be harnessed. According to IRENA, Cuba has a good potential in both solar and wind resources with an average solar irradiance of 223.8 W/m^2 (5.4 kWh/m^2/day) and average wind speed at around 5.7 m/s, and in the southeast above 7 m/s [8,9].

The Cuban government estimates that $3.5–4.0 billion in investments is needed to achieve their 2030 renewable energy targets with a significant share of foreign direct investments. The investments are foreseen in the wind and solar photovoltaic production. However, the government promotes investments in other renewable energy sources such as biogas, forestry biomass, agro-industrial residues and municipal solid waste [10].

Although transition to more renewable based energy systems are becoming more desirable, there are various technical challenges to overcome. Vazquez et al. argue that *"increased share of RES in future electrical power system brings several challenges to system planning and operation. Weather dependence of wind power and solar PV generation increases uncertainty in the premises of system design, which should be taken into account in decision making about required generation capacity and reserves, need of energy storages, control strategy and flexibility capacity of the system"* [11].

The study case, Isla de la Juventud, is the second largest island in Cuba, located in the southwestern part of the country, with an area of 2419 km^2 and a population of approximately 89,000 (35.73 persons/km^2). Isla de la Juventud has similar characteristics to the electric power system around Cuba although on a smaller scale, making it an ideal case to examine behavior of the radial electricity system with 100% distributed generation, made up of five main 34.5 kV circuits that supply energy to the distribution substations. The electrical power system under study is isolated from the National Electrical Power System (NEPS), operating autonomously. The generation of the system is made up by

11 diesel and fuel oil generators with an installed capacity of 35.44 MW and three solar parks (La Fe (0.8 MW), Universidad (2.4 MW) and Los Colonos (1 MW)) with 4.2 MW; a biomass plant (La Melvis) with 0.5 MW and one wind farm (Los Canarreos) with 1.65 MW of capacity. Currently 16% of installed capacity is from RES. The system is made up of fossil fuel-based generators with installed capacities up to 3.9 MW each, with similar maximum and minimum active power, power factor, ramp rates and fuel consumption in g/kWh (four MAN generators with capacity of 3.85 MW each, four BAZAN generators with a capacity of 3.6 MW each and three MTU generators with a capacity of 1.88 MW each). For this reason, the load share served by each generator is quite similar, and any of these generators, which act as the base load generation, can be used in the normal operation of the power system. The MAN type generators represent the basic generation system, with the BAZAN type for the reserve and MTU generators supporting the maximum peaks [12,13].

A desk review of RES utilization on islands globally shows that different approaches and tools have been employed e.g., to evaluate the situation of the existing electricity generation mix [14,15], to analyze the potential of energy efficiency to reduce electricity demand [16], to determine the existing barriers to the RES projects considering financial, and institutional, social or political aspects [17].

An implementation of long-range development analysis is crucial to achieving Cuba's energy and climate goals. The long-term planning analysis identifies overall transmission needs for a future timeframe, given demand growth, the targeted energy mix, interconnection policies and RES locations, among other factors [18]. In particular, alternative forecasting with existing renewable energy potentials, economic and technological variables is needed to decide the best alternatives. Long-term energy planning models are used to define investment paths and to inform long-term strategic decision making over the development of a national energy system. Long term planning models and tools have been used widely for generation expansion planning with a long (15–40+ years) planning horizon [19].

Internationally, a wide range of diverse energy planning tools are available based on the objectives they fulfil, the technologies they consider, and the time-steps they analyze. Connolly et al. suggest that to generate a long-term 'storyline' for implementing 100% renewable energy-systems, Invert simulation tool, EnergyPlan and the Low Emissions Analysis Platform (LEAP) may be the most suitable due to their lengthy scenario-timeframe [19]. Similarly, according to IRENA tools such as e.g., MESSAGE, TIMES, MARKAL, OSeMOSYS, WASP and BALMORE can be suitable tools for 20–40 year timescale in similar analysis [18].

IRENA found that most developing and emerging economies suffer from a lack of data availability and technical know-how that pose serious challenges to focus on ensuring solid capacity, flexibility, transmission capacity and—in certain contexts—stability, which can also compromise the use certain tools. In the end, there are no ideal tools to suit all purposes but data availability, specific objectives and purpose of the study and the conditions of the site define the choice of the right tool [18]. In this article the criteria in selecting the most appropriate tool considered the accessibility to the tool (free open access), the type of tool, future orientation, and previous studies carried out with the tool. Thus, this research utilizes the Long-range Integrated Development Analysis (LINDA) model to project future scenarios for Isla de la Juventud. LINDA is an Excel-based tool used for energy systems analysis and building future scenarios. It has been used to model future energy systems e.g., in Cambodia and Lao PDR [20], Thailand [21], China [22], Barbados [23] and Cuba [24].

This article provides the technical basis for integrated development roadmap analysis that helps Cuba to achieve best RES penetration composition. We analyze two alternative scenarios for future development in accordance with existing renewable energy potentials and technological variables with Isla de la Juventud as a case study. The results can be applied to model provincial and national level power electric systems as Isla de la Juventud has similar characteristics to the electric power system around Cuba. The analysis can provide direction on how the Cuban national system would behave with high levels of renewable energy sources integration, and point out solutions for different shares of RES in the national grid in the future.

2. Materials and Methods

This article utilizes The LINDA (Long-range Integrated Development Analysis) model which is based on intensity approach, building on the Extended Kaya Identity, which is used for the calculation of CO_2 emissions as depicted in equation below:

$$CO_2 = \frac{CO_2}{TPES} \times \frac{TPES}{FEC} \times \frac{FEC}{GDP} \times \frac{GDP}{POP} \times POP \quad (1)$$

where,

- CO_2 is carbon dioxide emissions (e.g., ton, kton) from fuel combustion;
- TPES is total primary energy supply (e.g., ktoe) (including all fuels and other forms of primary energy, before the combustion process and transfer and distribution of electricity or heat);
- FEC is final energy consumption (e.g., ktoe), meaning consumption of energy carriers such as district heat and electricity, and fuels used in residential heating and transport;
- GDP is gross domestic product in real prices (e.g., USD); and
- POP is the amount of population (e.g., person).

LINDA is a so-called 'Accounting Framework' type of model which allows the user to construct various economic scenarios by choosing different economic growth rates for different sectors including agriculture, industry, transportation and services as shown in Figure 1. Here, the energy use is divided into fuels and electricity, with energy intensity defining how much economic output is generated with a certain amount of energy used. Economic structures will affect the energy demand as intensities differ by economic sectors. By changing the energy intensities in the scenarios, the user can have an impact on the final energy demand. The energy intensity of a sector can decrease due to the introduction of a more efficient technology or shift to less energy intensive products or production structure [20,22,24].

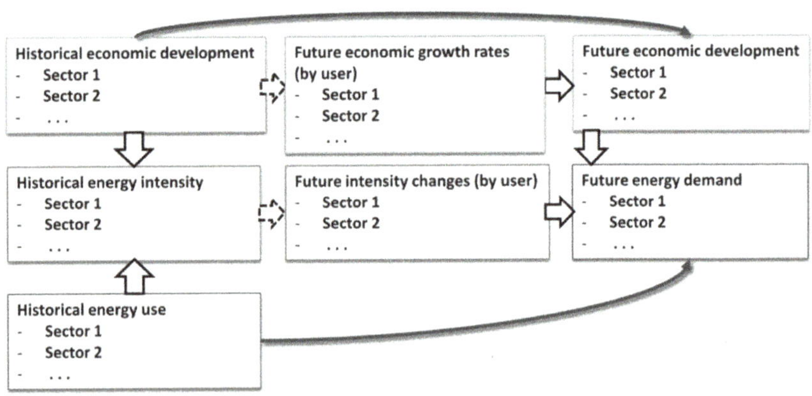

Figure 1. Calculation procedures based on the historical development and user inputs to define future energy demand [20].

The scenario construction process with the LINDA model starts with the decision of annual future economic growth level for different sectors and the future changes in the sectoral energy intensities. These provide data for annual future energy demand in different sectors. The load curve and its future changes for different consumer sectors for weekdays and weekends as well as different months are given to construct hourly consumption scenarios based on the yearly demand data. The yearly investments in electricity production capacity by power plant type are given and the model balances the production and consumption every hour by calculating the supply from variable renewable sources (wind and solar) and subtracting this from the total demand to get the residual load which is produced with the other power plants based on their given priority order. The model calculates the CO_2 emissions based on the characteristics of different fuel and the used amounts. The calculation linkages between different modules are shown above in Figure 2.

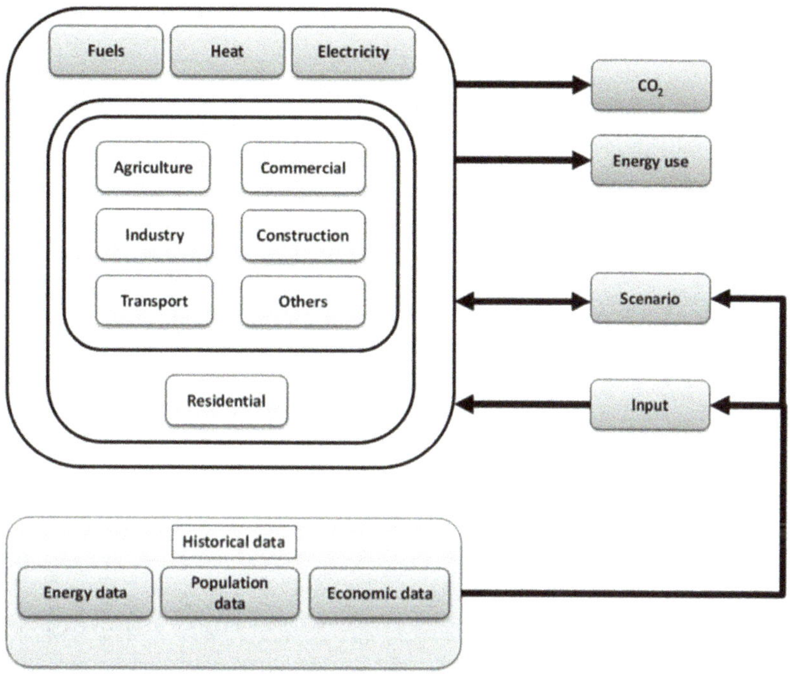

Figure 2. Calculation linkages between different modules in the LINDA model [20].

The data used in the modeling for the scenario analysis are taken from the International Energy Agency (IEA) World Energy Statistics [25], National Statistics Office of Cuba (ONEI) [26] and the electric company of Isla de la Juventud [27]. For the percentages and estimates on future growth, experts from the UNE were consulted, who provided sensitive information on investments that would be made in the Isla de la Juventud in renewable sources, as well as the real load curve of the power system. The information was further processed to create an annual load curve for 2019 and cross checked with published data on the ONEI website. The historical data from ONEI provides statistical information on all sectors, divided by provinces. From the classification of the obtained information, the LINDA model allows a sectoral analysis on:

- GDP based on industry, agriculture and forestry, transportation and commercial sectors; and
- Electricity use in industrial, agriculture and forestry, transportation and residential sectors.

On the growth projections the authors defined sectoral growth rates for the future based on the historical data available and in relation to the projections of the electricity demand growth and electricity intensity. The LINDA model utilizes hourly load curves for different sectors of the economy to analyze future sectoral and total electricity demand. The model user inputs the hourly load curves for weekdays and weekends and for different months for one year for different sectors of the economy as well as future projected load curves for all the future years of the scenario. The estimations of the future growth in electricity consumption are based on the views of experts including the UNE. Figures 3 and 4 illustrate the examples of a typical weekday in January in household and commercial sectors.

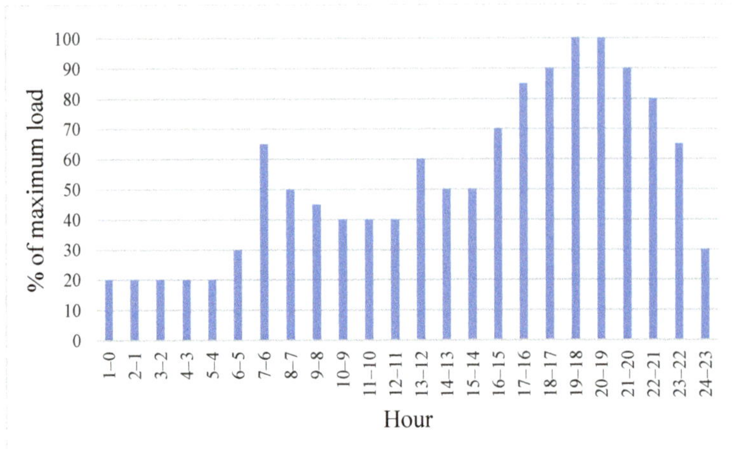

Figure 3. Load curve for household electricity consumption for a weekday in January 2019 (percentage of the sectoral maximum load). Source: own elaboration with the model [27].

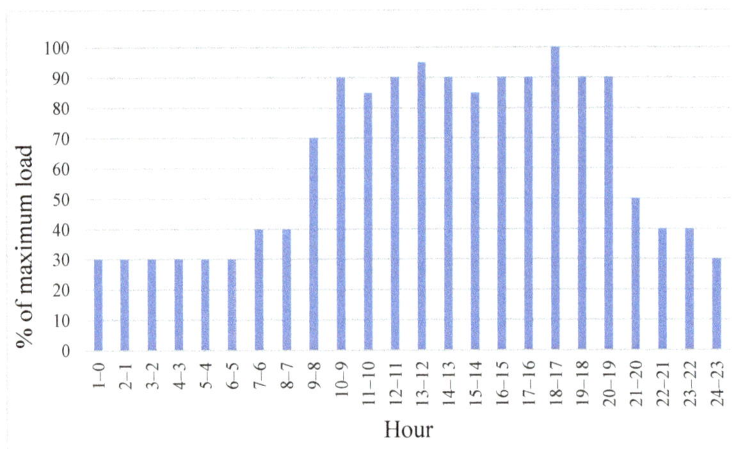

Figure 4. Load curve for commercial sector electricity consumption for a weekday in January 2019 (percentage of the sectoral maximum load). Source: own elaboration with the model [27].

Data on solar radiation and wind are obtained from The Modern-Era Retrospective analysis version 2 (MERRA 2) databases [28,29] and are shown in Figures 5 and 6.

In the electrical power system, the demand and supply have to be in the balance every hour of the year. LINDA calculates the electricity demand for every hour of the year and matches supply with the demand. The residual load is first calculated for different types of production. This residual load is the hourly demand minus the hourly production by the intermittent renewable energy sources, in this case, wind and solar. The calculation is illustrated in the below equation.

$$RL = D - G \quad (2)$$

where,

- The residual load (RL);
- Demand(D); and
- Intermittent energies renewable generation (G).

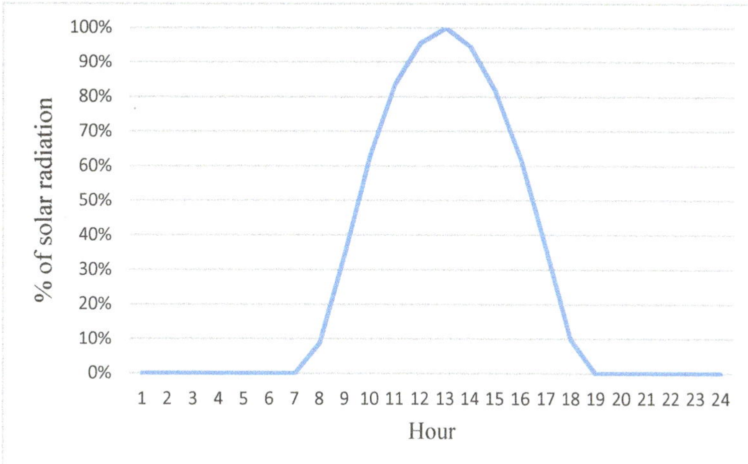

Figure 5. Hourly solar radiation curve for one day in Isla de la Juventud. Source: own elaboration with the model [28].

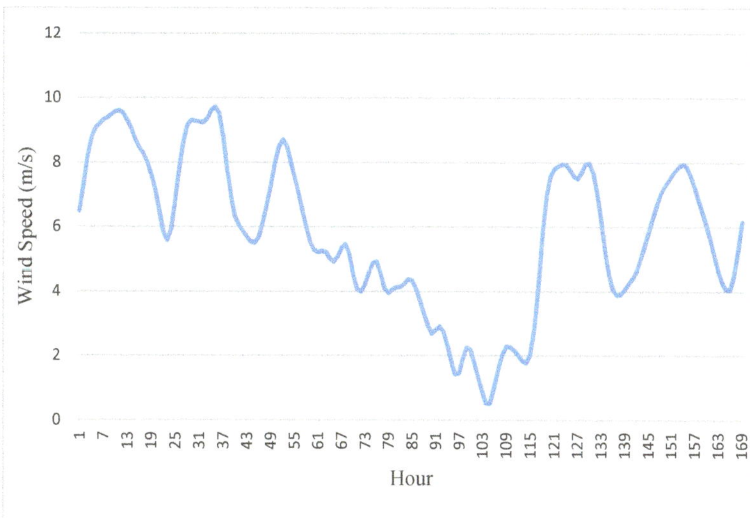

Figure 6. Hourly wind speed measurement (m/s) for one week in "Isla de la Juventud" (50 m height). Source: own elaboration with the model [29].

The production of the residual load has to be carried out with power plants that can be controlled such as fossil fuel condensing power plants, diesel power plants, biomass power plants or hydro power plants (or using storage if it exists). In Cuba, the hydro capacity is so small that it cannot solve the problem of grid balancing.

LINDA allocates the residual load for the fossil fuel based power plants and the user can give them a priority order to define which type of power plants produce and how big share of the residual load. In addition, the model calculates the required ramping rate for residual load production. The ramping rate is calculated for the required 1 to 6 h maximum (increase) and minimum (decrease) changes in residual power production as well as average increase and decrease of residual power production.

3. Results

The results are organized into sections to introduce the current situation: base year situation in 2019 in 3.1, renewable energy scenario (RENES) in 3.2, 100% renewable energy scenario with (MAXRES) in 3.3 and a comparison of results in 3.4. The authors used the data of the historical economic growth rates of Isla de la Juventud [25,26]. The inputs introduced in the model consider the country's policies to achieve a 30% penetration (of installed capacity, 24% of electricity generation) of renewable sources by 2030 [6] and a total growth of user defined GDP growth of 11.7% until 2030. For the different scenarios we have assumed that the installed capacities for RES increase in a renewable scenario (RENES) for solar PV up to 19 MW (34% of the total installed capacity) with biomass remaining the same (1% of the total installed capacity) and in the maximum use of renewable sources (MAXRES) scenario solar PV is increased up to 19 MW (31% of the total installed capacity), wind to 6.65 MW (11% of the total installed capacity) combined with a fuel switch from diesel and heavy-fuel oil to biofuels up to 35.94 MW (58% of the total installed capacity). Furthermore, in both scenarios, 10 MW in batteries are installed to store excess energy.

3.1. Base Year (2019)

The LINDA model utilizes the information provided by the national statistics office in its annual summary of the development of each province and its historical rates of economic growth. [26] The historical data from 2012 to 2019 is used for each sector of the economy.

Figures 7 and 8 show the economic growth and historical energy use for the different sectors for the period from 2012 to 2019. The gradual growth can be seen in both the value added and electricity consumption. The analysis indicates that the residential sector is historically the largest consumer in the system under study [26].

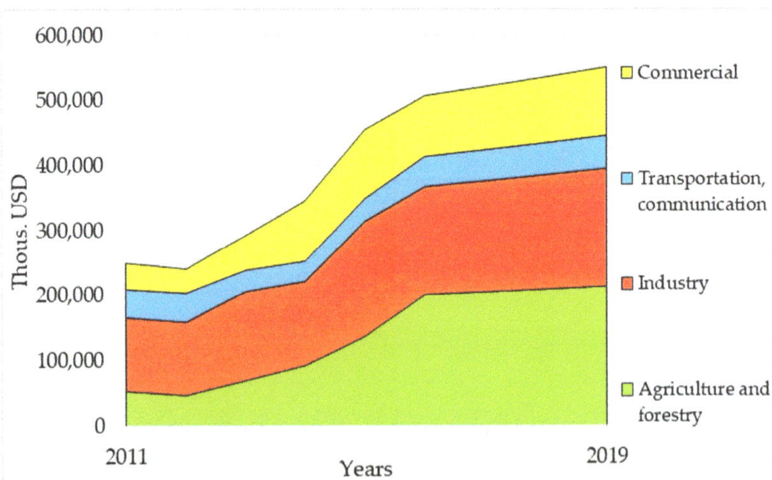

Figure 7. Historical data for Value added (GDP) in Isla de la Juventud (2011–2019) Source: own elaboration with the model [27].

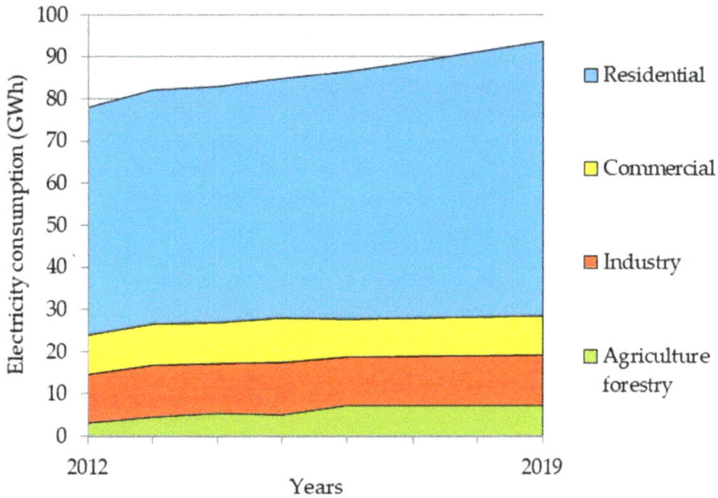

Figure 8. Electricity consumption by sector in Isla de Juventud in 2012–2019. Source: own elaboration with the model [27].

This is due to the low activity in the service, industry and agriculture sectors, similar to the structural behavior on the main island of Cuba. As electricity production is dominated by generators fueled by petroleum products as primary energy sources (Figure 9) in 2019, the base year, total CO_2 emissions were 82 Mt.

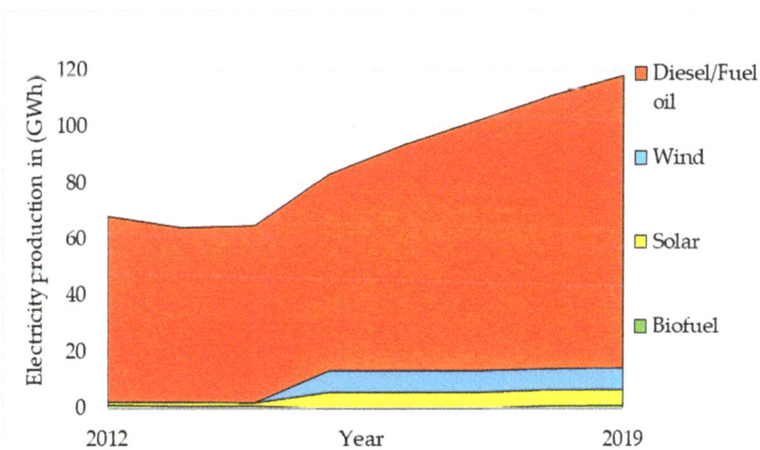

Figure 9. Electricity production in Isla de Juventud in 2012–2019. Source: own elaboration with the model [27].

Figure 10 shows the behavior of a typical winter day versus a summer day, showing that the system is predominantly residential with similar characteristics to the Cuban electro-energetic system. The main characteristics of the system under study reveal the following details:

- Residential sector electricity consumption defines the overall load profile of the total demand with maximum peak taking place in the evening from 18:00–22:00 during both winter and summer season.

- The peak in the summer occurs around 22: 00, reaching around 20 MW
- The peak in winter occurs around 19:00, reaching around 18 MW
- In the summer curve, the difference between maximum and minimum consumption is around 5 MW, and the load is more uniformly balanced during the day
- In the winter curve the difference between the maximum (peak) and minimum consumption is around 9 MW, showing larger variation in the load profile.
- When comparing the winter and summer curves there is a difference of around 10 MW between the summer peak and the winter minimum
- These aspects are crucial in planning demand and supply, as well as the reserve capacity to perform a cost effective economic operation of the system, without jeopardizing the stability of the electric power system.

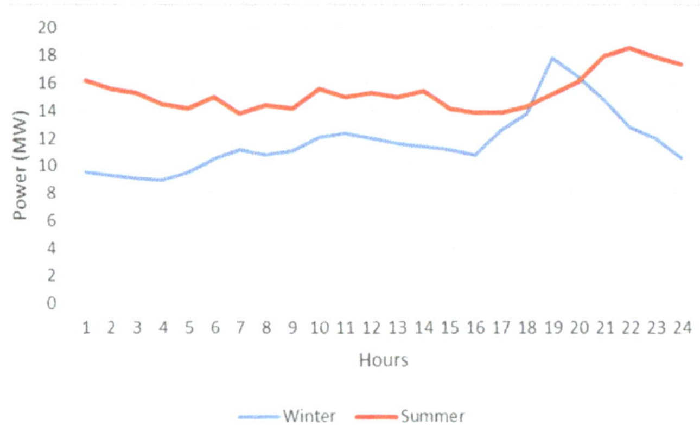

Figure 10. Demand behavior curves for a typical winter/summer day in 2019. Source: own elaboration with the model [27].

Figure 11 shows total monthly energy consumption in Isla de la Juventud in 2019. [27]. It can be seen that summer months, mainly July and August, are the months with the highest energy consumption, with over 10,000 MWh. In the winter months, this consumption decreases considerably, with just over 7000 MWh consumed in February. In February, electricity consumption is 36% less than in July. The annual demand of the system under study in 2019 was 114,548 MWh [27].

Figure 11. Total electricity demand on Isla de la Juventud in different month of 2019 Source: own elaboration with the model [27].

In Isla de la Juventud, by the end of 2019, there was around 10–13% of penetration of renewable energy into the electrical system. Figure 12 shows the contribution to electricity generation by different renewable energy sources, mainly wind and solar (biomass use for electricity production is minimal) and the total penetration in 2019. The results show a maximum penetration of just under 14% in total for all the RES. During the months from February to April RES provide a larger share of the energy of the total demand because of more solar production. February has the highest RES penetration rate, covering more than 13% of electricity demand. Solar energy has the highest penetration in the months from February to April with more than 8%, and at least 6% in all months of the year.

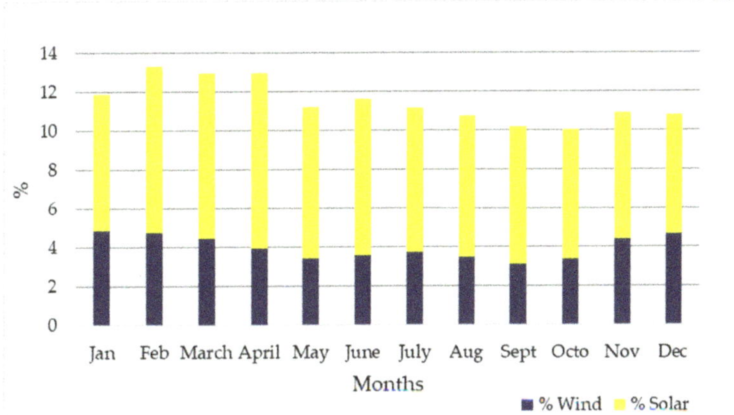

Figure 12. Renewable energy penetration rate (%) in 2019. Source: own elaboration with the model [27–29].

The share of wind energy does not exceed 5% of monthly consumption. During the months from August to October, wind power production has the lowest share and from December to February, the penetration is higher.

An important aspect for the analysis is to compare the penetration of renewable energy sources and hourly demand with the residual load curve. The behavior of the system has been analyzed against the influence of variable renewable sources. From the point of view of operation, it is necessary to observe the part of consumption to be covered by conventional generation each hour of the year. Figure 13 illustrates the demand versus residual load on a typical summer day. A further analysis shows that the greatest influence of renewable sources occurs during 9:00–16:00, mainly due to solar energy production. The biggest difference between the load curve and the residual load curve is at midday due to the peak of solar production. The difference is smaller at night and at dawn, because solar energy production reaches zero and only wind energy contributes to the generation and is proportionally reflected on the residual load curve. The difference between the load curve and the residual load curve is around 6 MW at maximum.

3.2. Modeling Renewable Scenario Analysis (RENES) in 2030

This scenario is based on the historical growth rates of previous years and the assumed future growth rates. The installed capacity in solar energy increases to 19.2 MW (34% of total installed capacity), wind power capacity remains at 1.65 MW (3% of total installed capacity), as well as 0.5 MW (1% of total installed capacity) of biomass. The renewable share reaches 38% of the total installed capacity by the year 2030. The assumptions in increase of solar and wind reflect the government plans and user defined inputs. With the assumed GDP growth of 11.7% in the different sectors the final energy consumption of the residential sector is expected growth by 10.5%, and industry and commerce, by 8% and 7.5% respectively.

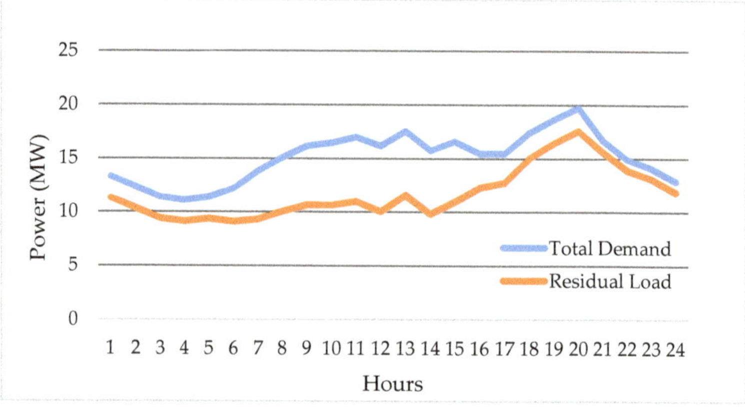

Figure 13. Residual load curve vs. load curve in 2019 for an example day. Source: own elaboration with the model [27].

The growth behavior is shown in Figure 14 with the residential sector clearly dominating. Similarly, in Figure 15 we can see the dominance of residential sector in the electricity consumption. Figure 16 shows that despite the increase in the installed capacities of RES into the system under study, fossil fuel consumption dominates the electricity production for the period from 2015 to 2030. These results show that the electric power system remains highly dependent on fossil fuels in the scenario with RES production at covering around 25% of the total electricity generated in a year as can be seen in Figure 17. The greatest contribution from renewable energy sources can be observed from February to April, as solar power capacity increases, its contribution grows considerably, with monthly penetration values of over 30%, and an annual average penetration of 28.5%. These values are in line with the Cuban energy sector targets 24% by 2030.

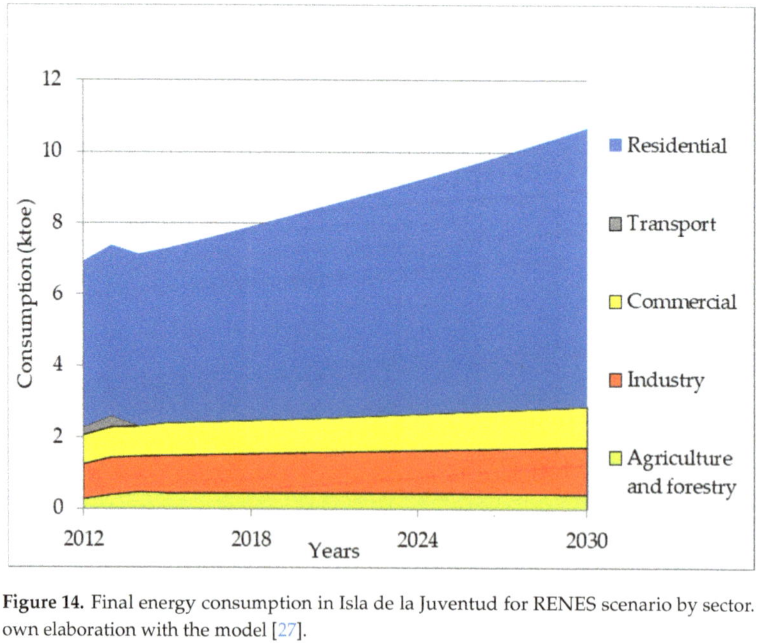

Figure 14. Final energy consumption in Isla de la Juventud for RENES scenario by sector. Source: own elaboration with the model [27].

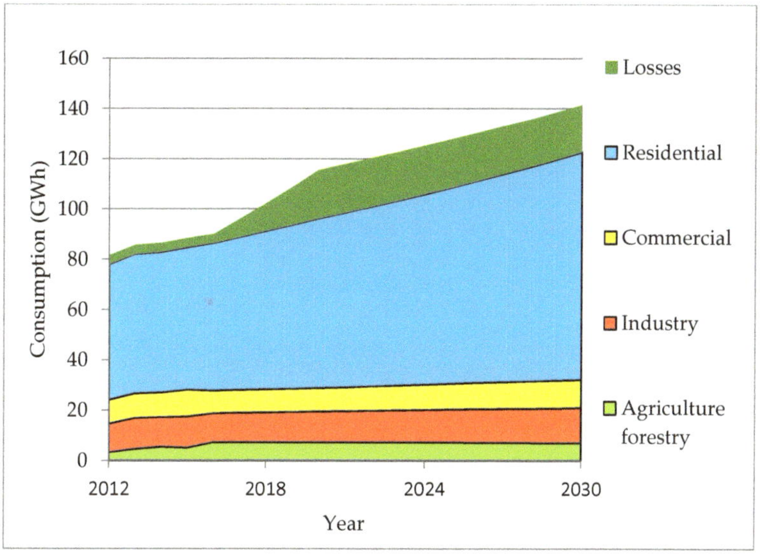

Figure 15. Electricity consumption of Isla de la Juventud in 2030 in the RENES scenario. Source: own elaboration with the model [27].

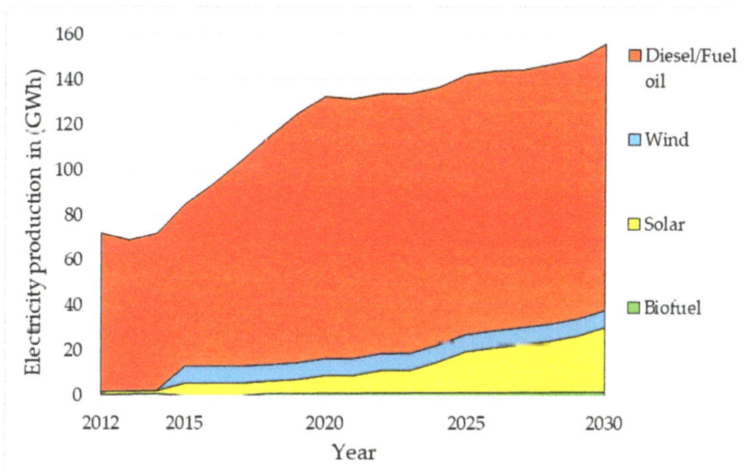

Figure 16. RENES scenario for electricity production until 2030. Source: own elaboration with the model [27].

Figure 18 shows coverage of demand by different energy sources for one day in April and Figure 19 one week in April 2030. The contribution from solar is highest during the day hours while in the early mornings and night time the most contribution is from diesel. Solar energy reaches a maximum of about 12 MW during the midday hours and wind energy with a maximum occurring mainly at night and early in the morning.

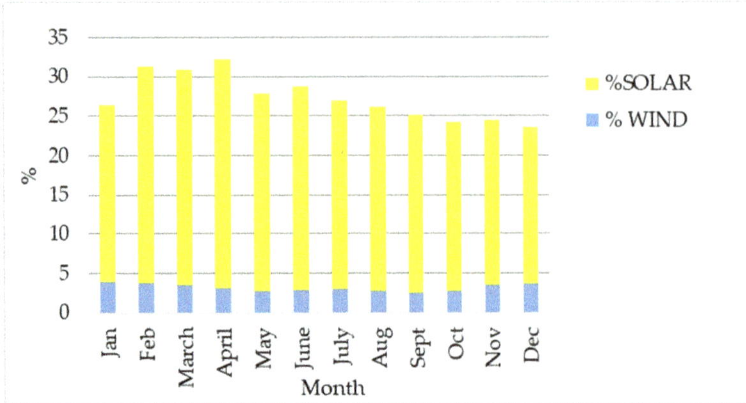

Figure 17. Renewable energy penetration rate (%) in different months in the RENES scenario. Source: own elaboration with the model [27–29].

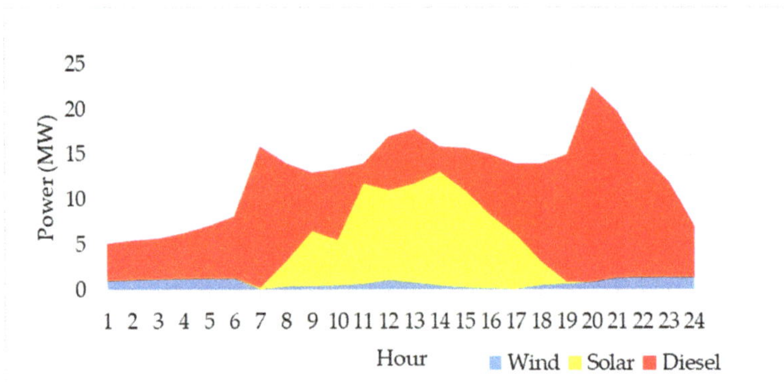

Figure 18. Electricity production for a day (19 April 2030) for the RENES scenario. Source: own elaboration with the model.

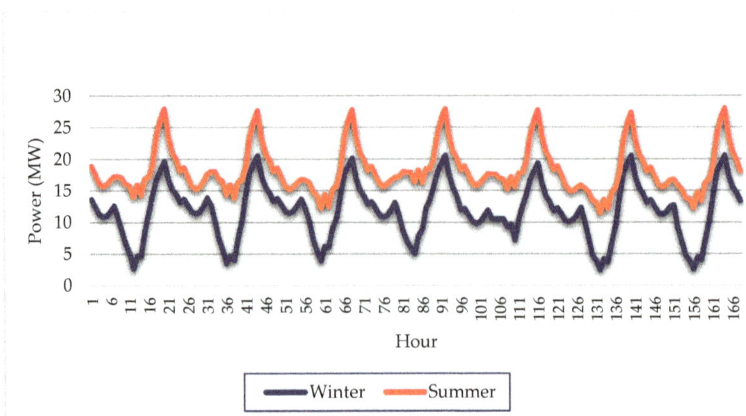

Figure 19. Residual load curve for a typical winter week versus summer week in 2030 for the RENES scenario. Source: own elaboration with the model.

As shown in Figure 17 the winter months have the highest contribution of RES to the system; during the winter months there is an excess of solar production which could be stored in battery systems. On the contrary, in the summer months the consumption is higher and the contribution of the RES cannot cover the demand. In Figure 19 we can observe the residual load during a winter and summer week in 2030, showing the residual load to be less in the winter and more in the summer. The residual load shows that the batteries would be an alternative to take advantage of the hours of maximum solar production by storing the energy to give the electrical power system a backup during the hours of maximum generation

The introduction of batteries as shown in Figure 20 would increase the installed capacity in the system under study by 10 MW and therefore the penetration of the RES in this simulation would reach over 40% giving the system more independence. As solar energy generation capacity increases, it is noticeable that the demand for fuels (oil and its derivatives) decreases by 5% in the period from 2020 to 2030. Figure 21 shows the CO_2 emissions to the atmosphere from 2020 to 2030. Firstly, an increase in emissions is observed, due to the increase in demand and the use of diesel and fuel generators to cover this increase. However, after installing the 19 MW of solar energy according to the plan, a decrease in emissions is observed, reaching 78.2 Mtons in 2030.

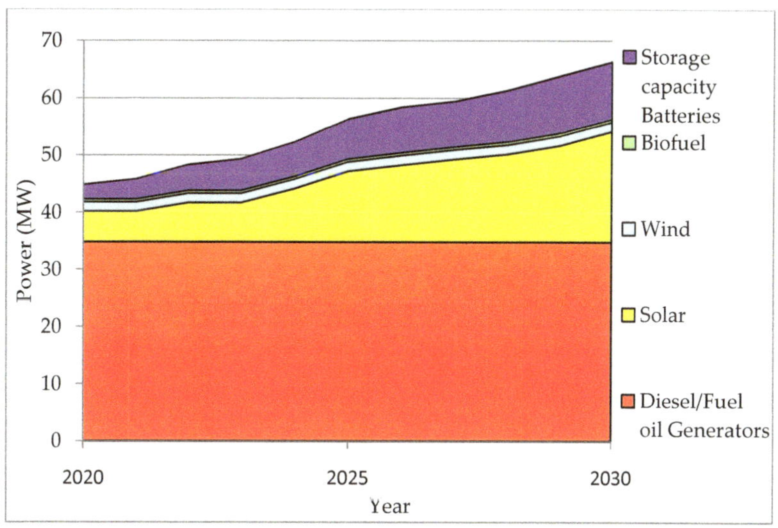

Figure 20. Capacity of power plants in Isla de la Juventud for the RENES scenario with storage batteries. Source: own elaboration with the model.

3.3. *Modeling Scenario for Maximizing the Use of Renewable Energy Sources (MAXRES) by 2030*

In this scenario we maintain the growth rates of RENES and increase the solar capacity to 19.2 MW, wind capacity up to 6.2 MW and 10 MW in energy storage. In addition, fuel switch from fuel oil and diesel to biofuels is realized with a total of 35.94 MW of biofuel generators. This significantly increases the penetration of renewable energy sources gradually up to 100% in electricity generation on Isla de la Juventud with the fuel switch from a non-renewable to renewable fuels (Figures 22 and 23).

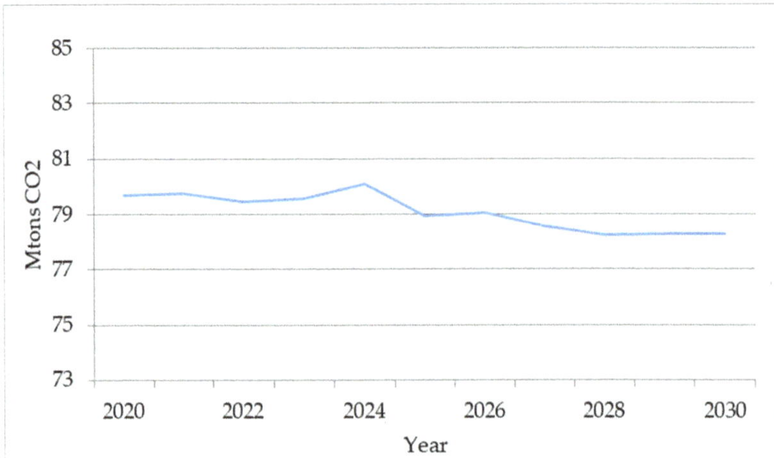

Figure 21. Total CO_2 emissions in the RENES scenario. Source: own elaboration with the model.

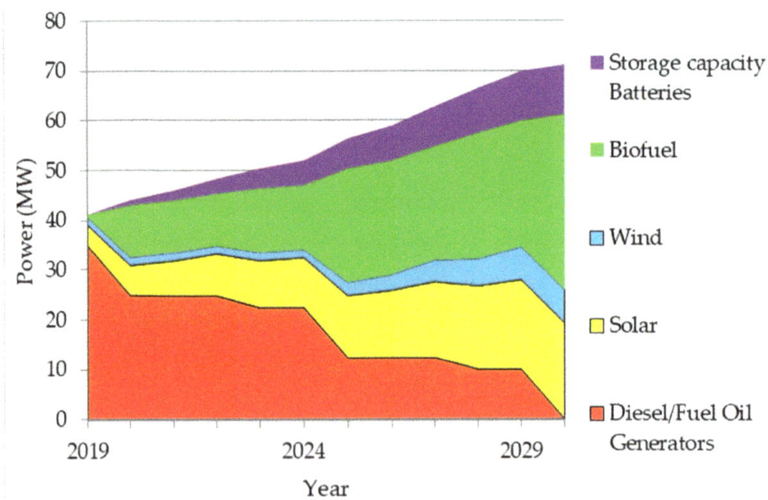

Figure 22. Power plant capacity on Isla de la Juventud for the MAXRES scenario. Source: own elaboration with the model.

In this scenario, the newly installed wind energy capacity results in an increase of 50 GWh annually more than in the RENES scenario. This extra energy, added to that of solar energy, is stored in the batteries to be used as a backup in case of emergency or during peak demand hours, thus avoiding generation losses in the system.

With the transformation of the energy matrix to 100% RES based for 2030, the amounts of CO_2 emitted to the atmosphere in this period gradually achieve a 100% reduction of emissions as shown in Figure 24.

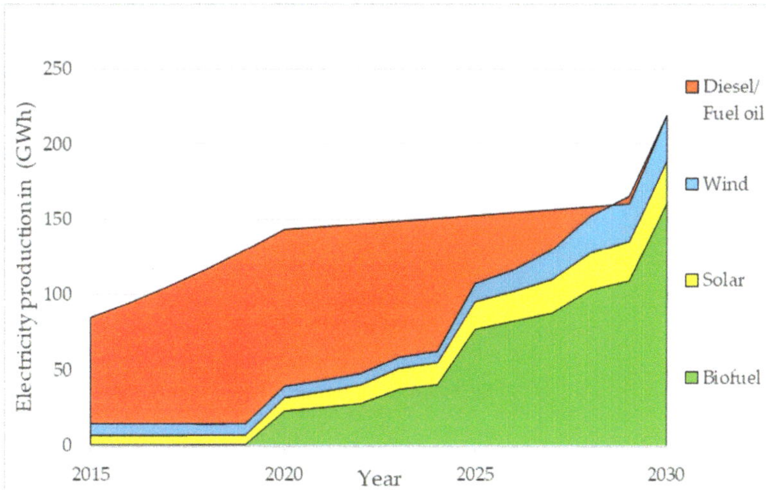

Figure 23. Electricity production on the MAXRES scenario. Source: own elaboration with the model.

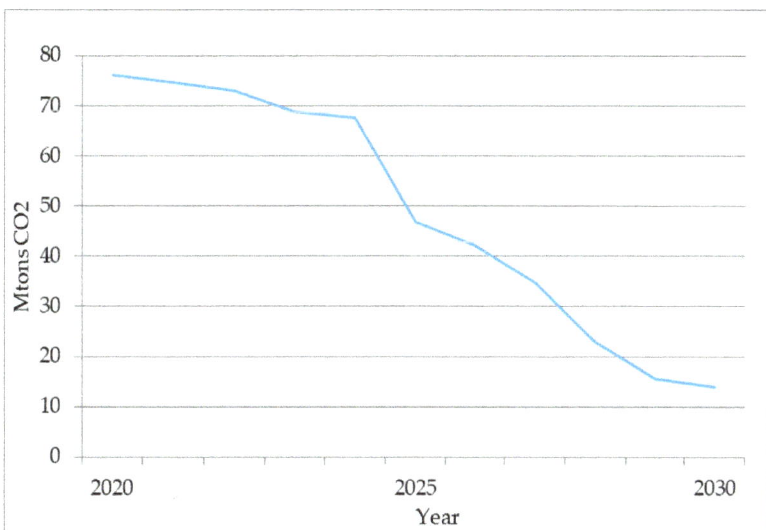

Figure 24. CO_2 emissions in the MAXRES scenario. Source: own elaboration with the model.

3.4. Comparison of the Scenarios

The results show that both the RENES and MAXRES scenario comply with the country's energy policy targets for 2030, reaching a minimum of 30% of renewable sources in the total installed capacity. The residual load shows that in RENES scenario, photovoltaic solar energy makes the greatest contribution during the midday hours with the possibility to be used as a backup with the battery storage. Moreover, in the case of MAXRES scenario wind complements the demand requirements with biofuels and provides additional potential for storage especially during the night time. In terms of CO_2 emission from electricity generation, we see the gradual decrease from RENES and a reduction to 13.9 in the MAXRES scenario compared to base year.

The comparison of results in different scenarios is shown in Table 1 based on:

(1) Growth in GDP by the year 2030;
(2) The installed capacities of renewable energy sources are shown for the different scenarios, with base year 6.35 MW, the RENES 21.35 MW and a 10 MW Battery storage and MAXRES 61.79 MW and a battery storage of 10 MW;
(3) Percentage of RES of total installed capacity with base year accounting to 16%, the RENES to 43% and MAXRES to 100% of utilization of RES accordingly;
(4) Influence of the residual load
(5) CO_2 emissions; 82 Mtons in base year 78.2 Mtons in RENES scenario and 13.9 Mtons of CO_2, in MAXRES scenario respectively
(6) Electricity production: With the energy mix changes from one scenario to another and increases in installed capacity electricity production results in RENES scenarios at 155 GWh and MAXRES scenario at 210 GWh;
(7) Electricity consumption: in the base year at 95 GWh, RENES scenario at 140 GWh and MAXRES scenario at 160 GWh

Table 1. Scenario inputs and results.

Scenarios		Base Year	RENES	MAXRES
GDP Growth (%) by 2030		-	11.7%	11.7%
Installed capacity by RES	Solar	4.2 MW	19.2 MW	19.2 MW
	Wind	1.65 MW	1.65 MW	6.65 MW
	Biofuel	0.5 MW	0.5 MW	35.94 MW
	Batteries	-	10 MW	10 MW
% RES of installed capacity		16%	38%	100%
Residual load		Low influence	More solar input with battery backup	The greatest contribution is solar and wind with battery backup
CO_2 emissions		82 Mton	78.2 Mton	13.9 Mton
Electricity production		120 GWh	155 GWh	210 GWh
Electricity consumption		95 GWh	140 GWh	160 GWh

4. Discussion

The objective of this study was to question the dependence on fossil fuels in the future electricity system in Cuba. Isla de la Juventud provides an excellent case study for power system modeling due to its size and because it is an island system with similar characteristics to the Cuban main island power system.

The Cuban government has applied a very stimulating strategy to deployment his policy to development RES within the country: (1) to open the foreign investment to achieve RES technologies and financial capital, even with 100% of the foreign investment, (2) to increase the interaction with international organizations in order to develop "absorptive capacity" on RES and technologies (3) to give autonomy for the local government to decide their development strategy for increasing penetration with RES, (4) to develop a program to prepare the local governments to be involved with RES, (5) to involve universities with the local governments in order to enhance opportunities for different stakeholders to support the use of RES in local conditions, (6) to incentive private sector using RES. Consequently, the government needs to identify and raise awareness of key policy questions and their implications for the long term development of the Cuban electric power systems means "to quantitatively calculate the direction of future energy policy and the implications of taking one pathway of energy sector development instead of others" [30].

It has been recognized in literature that desirable generation expansion scenarios or renewable energy targets depend mostly on policy priorities and on economic resources, rather than on technical grounds [15,16,18,31,32]. At the same time, the particular characteristics of a given isolated power system influences its capability to safely integrate increasing shares of RES. Although this analysis is mainly based on planning alternatives, some technical and operational aspects of the system were considered for the selection of the scenarios. These include aspects such as generation capacity and technology, maximums and minimums of power by conventional generators, ramp rate, load share, generation dispatch, grid congestion and stability issues.

The system is made up of fuel generators with installed capacities up to 3.9 MW each, with similar maximum and minimum active power, power factor, ramp rates and fuel consumption in g/kWh. For this reason, the share of load served by each generator is quite similar, and any of these generators, which act as the base load generation, are used in the normal operation of the power system. It is important to highlight that even when flexibility analysis for stability control and other parameters need further analysis, some stability and operational conditions were considered. A stability study carried out in the power system of Isla de la Juventud (by the authors) consider certain operating conditions that should not be violated under any circumstances, although these results are sensitive and are not described in this article, they served as authors criteria constraint to reject certain results that did not meet these conditions. Thus, only the technically valid scenarios for the local conditions were prioritized in the analysis.

As a particular characteristic of the electric power system of Isla de la Juventud, residential sector is the largest consumer in the system due to the low activity in the service, industry and agriculture sectors. Consequently, today the peak demand is about 20 MW around 22:00 during a summer day, exceeding the demand for a winter day by 2 MW. In Isla de la Juventud, by the end of 2019, there was 15% of penetration by renewable energy into the electrical system, with a significant participation of PV (66.14% of RES) [27]. Until 2030, UNE has projected an increase with PV up to 15 MWp. Therefore, the difference between the load curve and the residual load curve is around 6 MW at maximum.

The analysis of the scenario RENES confirmed that the batteries would be an alternative to take advantage of the hours of maximum solar production by storing the energy to give the electrical power system a backup during the hours of maximum demand. Moreover, it is possible to increase the share of intermittent renewable power generation with the use of energy storages, such as electric batteries, as has been concluded in several studies of island isolated power system [33,34]. The results show the scope of RES introduction to the electric power system is technically possible and abundant solar, wind and availability to biomass can provide a transition up to 100% RES based systems, similarly to other islands in the Caribbean [35]. The MAXRES scenario supports the government plan to increase the use of biomass for electricity production as Cuba has plans on promoting biofuel production nationally [7].

The growing concern about global climate change also drives the transition to non-fossil-based electricity production and the study shows that with increased RES the potential for climate mitigation is realized. The advantages of biomass utilization in the energy sector also support Cuba's sustainable development goals as stated by e.g., Bravo Hidalgo [36–38]. Similarly, decreasing costs, especially for wind and solar power, now offer a competitive alternative to conventional energy sources. Reducing the consumption of conventional sources towards zero decreases import dependence and supports sustainable energy transition necessary in Cuba to promote economic development [23].

The approach of the LINDA model enables the economic analysis of different scenarios in the view of future demand of the Isla de la Juventud power system and achieving different coverage with renewable sources. The LINDA model inputs include characteristics of the generation mix (also maximum and minimum required capacity of the fuel and diesel generators); sectoral economic development and sectoral energy intensity development; future growth rate for economy; future energy demand; installed and future power plant capacity and their respective load factors. The used Linda model version calculates the electricity consumption and production on an hourly basis and requires modifications if 15-min interval data is needed. The constructed model for Isla de la Juventud has only four economic sectors due to the data availability and can deal with structural changes within the industrial sector only by modifying the whole sector intensity and growth figures. In this case the data availability is the problem because more detailed subsector data was not available. The model itself can include as many sub-sectors as needed.

For the scenarios constructed, however, subsequent flexibility studies are necessary e.g., to study if the transmission lines of the power system have the necessary capacity to transmit the energy from renewable sources and to avoid possible blackouts or curtailments. In addition, a stability analysis linked to system operation analysis is important to guarantee voltage and frequency stability, either in normal operating conditions or in the occurrence of a fault. The priority order of the power plants for producing the residual load is not using optimisation algorithms in LINDA model because information of the ramping rates and ramping costs of the different power plants was not available. The LINDA model does not include the transmission and distribution network and hence the distribution costs and potential bottlenecks are neither considered in the modelling.

The temporal and spatial resolution of the applied LINDA model ensures an adequate quality and validity of the results compared with other models. The temporal resolution of LINDA limits its scope to the evaluation of hourly demand and generation balancing by the sectors of the economy. The evaluation of the results indicates the study findings are quite accurate to real situation currently, although the data is incomplete and at times estimations of the parameters were used due to lack of accurate data. The sensitivity analysis or optimisation could not be carried out comprehensively due to availability or confidentiality of the data. Unlike other similar studies with LINDA, e.g., [22–24] the optimisation (especially least cost options) studies was found challenging due to complex trade agreements, heavily subsidized oil products and other factors that may distort sensitivity analysis and finding the "correct optimisation" results in Cuba.

This study with a LINDA model does not fully reflect operational restrictions, because it is based on a simplified technology representation to determine generation scheduling and reserve sizing. More detailed models need to be applied to reinforce the results of this work. The authors suggest further analysis of e.g., Flextool to study the reliability, resilience and stability of the system, as the main challenges in integrating high shares of RES into the electrical system. Furthermore, the economic analysis of the lowest cost systems is an essential part of the analysis and equally important to the technological analysis of the modeling of the electric power system based on RES. This article focused on a set of technological possibilities, hence the authors recommend that further studies be carried out on the economics of the system on Isla de la Juventud. Only in this way will it be possible to identify the most suitable and lowest cost systems. An economic analysis of biofuel production was not carried out in this study, which provides another interesting area to examine the food-energy nexus: the costs and benefits of growing biomass for energy production versus food.

5. Conclusions

The aim of the article was to first see how Cuban policy goal with 24% penetration of renewable sources is possible. The electrical system of Isla de la Juventud has important similarities with the electrical system of Cuba and the economic, social and energy related data used by the model allowed analysis of different factors to showcase results that serve as an important evidence in the context of Cuba.

Secondly, geographical, economic and social conditions make Isla de la Juventud an ideal test scenario for Cuba, and the results of this work will help to understand the path that must be followed to achieve a power system with 100% of generation based on renewable energy, obtain maximum reduction on the use of fossil fuel and greenhouse gas emissions.

The use of the LINDA model allowed construction of different scenarios for the introduction of renewable sources in Isla de la Juventud. In the analyzed scenarios, solar energy (19.2 MW of installed capacity) coupled with energy storage systems play a fundamental role in meeting the demand and allowing up to 38% penetration from renewable sources in RENES reducing greenhouse gas emissions to 78.2 Mton of CO_2 equivalent. If wind energy and biofuel are added to the generation mix, it is possible to achieve energy independence (also from fossil fuels with 100% RES based electric system) and reduce the emission of greenhouse gases only 13.9 Mton CO_2 equivalent per year. Both scenarios show, in comparison with the base year 2019 of with 120 GWh of production and 82 Mton CO_2 equivalent, a significant decrease in the emissions especially as the future estimated production increases to 129% in RENES and 175% in MAXRES compared to base year.

These results show transitioning towards sustainable energy and electric power systems is evidently possible in Cuba with the results shown in RENES and MAXRES scenarios. The analysis of the electrical power system for the different sectors of the economy offers the possibility to model the energy mix in the long term as well as to see the impact on the emissions and the influence of new installed RES capacities in Isla de la Juventud. Here open source tools such as Long-range Integrated Development Analysis (LINDA) can provide critical information on the different development trajectories.

The current policy aims are within reach and even exceeding the targets is also well within realism. This, however, requires robust analysis regarding the technical and economic possibilities in Cuba. Until recently, access to resources, mainly equipment and financing, has by far been the most limiting factor. Furthermore, enabling aspects such as legal and regulatory framework, investment incentives, pricing mechanisms and motivation of stakeholders are crucially important in meeting the challenge in moving towards 100% RES based electricity system. Drafting a clear roadmap on how to realize the transition is thus a priority that should be addressed adequately.

Author Contributions: Conceptualization, E.A.A., M.K., A.S.F.; methodology, software and validation E.A.A., M.K., A.S.F., J.L.; formal analysis, E.A.A., M.K., J.L.; investigation, E.A.A., M.K., A.S.F., J.L.; resources, E.A.A., M.K., A.S.F., M.L.F.S.d.R., J.L.; data curation, E.A.A., M.K., A.S.F., J.L.; writing—original draft preparation, E.A.A., M.K.; writing—review and editing, E.A.A., M.K., A.S.F., M.L.F.S.d.R., J.L.; visualization, E.A.A., M.K.; supervision, A.S.F., M.L.F.S.d.R., R.A.P., J.L.; project administration, J.L.; funding acquisition, J.L. All authors have read and agreed to the published version of the manuscript.

Funding: The authors wish to thank the Academy of Finland for funding the IRIS project (Integration of renewable intermittent sources in the power system (research project number 320229).

Institutional Review Board Statement: Not applicable.

Informed Consent Statement: Not applicable.

Data Availability Statement: The data presented in this study are available on request from the corresponding author. The data are not publicly available due to privacy and sensitive nature of the data.

Conflicts of Interest: The authors declare no conflict of interest.

References

1. International Energy Agency (IEA). Global CO_2 Emissions in 2019. 2020. Available online: https://www.iea.org/reports/global-energy-co2-status-report-2019/emissions (accessed on 15 February 2021).
2. International Renewable Energy Agency (IRENA). *Renewable Power Generation Costs in 2019*; International Renewable Energy Agency (IRENA): Abu Dhabi, United Arab Emirates; ISBN 664-978-92-9260-244-4. 2020. Available online: https://www.irena.org/-/media/Files/IRENA/Agency/Publica-665tion/2020/Jun/IRENA_Power_Generation_Costs_2019.pdf (accessed on 15 February 2021).
3. United Nations. Department of Economic and Social Affairs (DESA). 2021. Available online: https://unstats.un.org/sdgs/report/2019/Goal-07/ (accessed on 15 February 2021).
4. International Energy Agency (IEA). SDG7: Data and Projections Access to Affordable, Reliable, Sustainable and Modern Energy for All. 2020. Available online: https://www.iea.org/reports/sdg7-data-and-projections/access-to-electricity (accessed on 17 February 2021).
5. United Nations. Transforming Our World: The 2030 Agenda for Sustainable Development. 2015. Available online: https://sdgs.un.org/2030agenda (accessed on 9 February 2021).

6. Periódico Granma. Lineamientos de la Política Económica y Social del Partido. April 2011. Available online: http://www.cubadebate.cu/noticias/2011/05/09/descargue-en-cubadebate-los-lineamientos-de-la-politica-economica-y-social-pdf/ (accessed on 9 February 2021).
7. Periódico Granma. Actualización de los Lineamientos de la Política Económica y Social del Partido Para el Período 2016–2021. July 2017. Available online: http://www.granma.cu/file/pdf/gaceta/Lineamientos%202016-2021%20Versi%C3%B3n%20Final.pdf (accessed on 9 February 2021).
8. Unión Eléctrica, UNE. DESARROLLO DEL SISTEMA ELÉCTRICO CUBANO. 2017. Available online: https://slideplayer.es/slide/11976486 (accessed on 9 February 2021).
9. Zhao, Y. Power Shift in Cuba: Seven Reasons to Watch the Renewable Energy Sector in the Post-Fidel and Trump Era. 2017. Renewable Energy World. Available online: https://www.renewableenergyworld.com/2017/02/10/power-shift-in-cuba-seven-reasons-to-watch-the-renewable-energy-sector-in-the-post-fidel-y-era-triunfo (accessed on 9 February 2021).
10. Panfil, M.D.-R. The Cuban Electric Grid: Lessons and Recommendations for Cuba's Electric Sector. 2017. Environmental Defense Fund. Available online: https://www.edf.org/sites/default/files/cuban-electric-grid.pdf (accessed on 9 February 2021).
11. Vazquez, L.; Hohmeyer, O.; Vilaragut, M.; Diaz, D.; Majanne, Y.; Castro, M.; Luukkanen, J. Energy System Planning towards Renewable Power System: Energy Matrix Change in Cuba by 2030. *IFAC-PapersOnLine* **2018**, *51*, 522–527.
12. Santos, A.F.; Fernandez, M.C.; Valerino, J. Análisis de Penetración Eólica por en el Sistema Híbrido Diesel-Eólico de la Isla de la Juventud. 2013. La Habana: XI Congreso Internacional de Alta Tensión y Aislamiento Eléctrico (ALTAE 2013).. Available online: https://www.researchgate.net/publication/307629851_ANALISIS_DE_PENETRACION_EOLICA_EN_EL_SISTEMA_HIBRIDO_DIESEL_-_EOLICO_DE_LA_ISLA_DE_LA_JUVENTUD/link/57cd9e3108ae057987aab2e7/download (accessed on 12 February 2021).
13. Santos, A.F.; Fernández, M.C. Influencia del Parque Eólico de Los Canarreos en el Sistema Eléctrico de la Isla de la Juventud. La Habana: 17 Convención Científica de Ingeniería y Arquitectura. Available online: www.researchgate.net/publication/307631026_INFLUENCIA_DEL_PARQUE_EOLICO_DE_LOS_CANARREOS_EN_EL_SISTEMA_ELECTRICO_DE_LA_ISLA_DE_LA_JUVENTUD (accessed on 9 February 2021).
14. Kuang, Y.; Zhang, Y.; Zhou, B.; Li, C.; Cao, Y.; Li, L.; Zeng, L. A review of renewable energy utilization in islands. *Renew. Sustain. Energy Rev.* **2016**, *59*, 504–513. [CrossRef]
15. Thushara, D.S.M.; Hornberger, G.M.; Baroud, H. Decision analysis to support the choice of a future power generation pathway for Sri Lanka. *Appl. Energy* **2019**, *240*, 680–697. [CrossRef]
16. McNeil, M.A.; Karali, N.; Letschert, V. Forecasting Indonesia's electricity load through 2030 and peak demand reductions from appliance and lighting efficiency. *Energy Sustain. Dev.* **2019**, *49*, 65–77. [CrossRef]
17. Weir, T. Renewable energy in the Pacific Islands: Its role and status. *Renew. Sustain. Energy Rev.* **2018**, *94*, 762–771, ISSN 1364-0321. [CrossRef]
18. International Renewable Energy Agency (IRENA). *Planning for the Renewable Future: Long-Term Modelling and Tools to Expand Variable Renewable Power in Emerging Economies*; International Renewable Energy Agency: Abu Dhabi, United Arab Emirates, 2017; ISBN 978-92-95111-06-6. Available online: https://www.irena.org/-/media/Files/IRENA/Agency/Publication/2017/IRENA_Planning_for_the_Renewable_Future_2017.pdf (accessed on 12 February 2021).
19. Connolly, D.; Lund, H.; Mathiesen, B.V.; Leahy, M. A review of computer tools for analyzing the integration of renewable energy into various energy systems. *Appl. Energy* **2009**. [CrossRef]
20. Luukkanen, J.; Akgun, O.; Kaivo-oja, J.; Korkeakoski, M.; Pasanen, T.; Panula-Ontto, J.; Vehmas, J. Long-run energy scenarios for Cambodia and Laos: Building an integrated techno-economic and environmental modelling framework for scenario analyses. *Energy* **2015**, *91*, 866–881. [CrossRef]
21. Karjalainen, J.; Käkönen, M.; Luukkanen, J.; Vehmas, J. *Energy Models and Scenarios in the Era of Climate Change*; University of Turku, Finland Futures Research Centre: Turku, Finland, 2014; ISBN 978-952-249-275-3. Available online: https://www.researchgate.net/publication/276119218_Energy_Models_and_Scenarios_in_the_Era_of_Climate_Change/link/5550ec5a08ae739bdb9202ca/download (accessed on 9 February 2021).
22. Luukkanen, J.; Panula-Ontto, J.; Vehmas, J.; Liyong, L.; Kaivo-oja, J.; Häyhä, L.; Auffermann, B. Structural change in Chinese economy: Impacts on energy use and CO_2 emissions in the period 2013–2030. *Technol. Forecast. Soc. Chang.* **2015**, *94*, 303–317. Available online: https://doi.org/10.1016/j.techfore.2014.10.016 (accessed on 9 February 2021). [CrossRef]
23. Moore, W.; Korkeakoski, M.; Luukkanen, J.; Alleyne, L.; Brown, N.; Chambers, T.; Evans, A. Identifying Inconsistencies in Long-Run Development Plans: The Case of Barbados' Vision for Energy Development. *SSRN Electron. J.* **2015**. [CrossRef]
24. Salazar, I.; Luukkanen, J.; Seisdedos, L.V.; Korkeakoski, M.; Vázquez, A.S.; Majanne, Y.; Fuentefria, A.S. ELECTRICITY SUPPLY WITH RENEWABLE ENERGY SOURCES AND THE CUBAN ELECTRICITY SYSTEM: CHALLENGES OF SUPPLY-DEMAND BALANCE. 2018. Available online: https://www.researchgate.net/publication/329424950_ELECTRICITY_SUPPLY_WITH_RENEWABLE_ENERGY_SOURCES_AND_THE_CUBAN_ELECTRICITY_SYSTEM_CHALLENGES_OF_SUPPLY-DEMAND_BALANCE (accessed on 10 January 2021).
25. International Energy Agency (IEA). World Energy Outlook, WEO. 2019. Available online: https://www.iea.org/reports/world-energy-outlook-2019 (accessed on 15 January 2021).
26. Oficina Nacional de Estadística e Información (ONEI). Anuario Estadístico de la Isla de la Juventud 2018. 2019. Available online: http://www.onei.gob.cu/node/14626 (accessed on 15 January 2021).

27. Alfonso, F.M.M.; Mena, P.A.S. ISLA DE LA JUVENTUD, SU SISTEMA ELECTRICO Y LA ASIMILACION DE LA GENERACION CON FUENTES DE ENERGIA RENOVABLES. 2018. 19 edición de la convenión de ciencia ingeniería y arquitectura. In Proceedings of the Conferencia Internacional de Tecnologías aplicadas a Redes Eléctricas Inteligentes (CITREI), Havana, Cuba, 26–28 November 2018.
28. Renewables.ninja. The Modern-Era Retrospective Analysis Version 2(MERRA-2). Solar PV (Point API)-21.691. 2021. -82.816-Version: 1.1 (using GSEE v0.3.1)-License: https://creativecommons.org/licenses/by-nc/4.0/. Available online: https://www.renewables.ninja/ (accessed on 20 March 2021).
29. Renewables.ninja. The Modern-Era Retrospective Analysis Version 2(MERRA-2). Wind (Point API)-21.691, -82.816-Version: 1.1-License: https://creativecommons.org/licenses/by-nc/4.0/-. Available online: https://www.renewables.ninja/ (accessed on 20 March 2021).
30. Comité Central del Partido Comunista de Cuba. Conceptualización del Modelo Económico y Social Cubano de desarrollo socialista. Plan Nacional de Desarrollo Económico y Social Hasta 2030: Propuesta de Visión de la Nación, ejes y Social Hasta 2030: Propuesta de Visión de la Nación, ejes y Sectores Estratégicos. 2017. Cuba: Granma. Available online: http://www.granma.cu/file/pdf/gaceta/%C3%BAltimo%20PDF%2032.pdf (accessed on 15 January 2021).
31. Owusu, P.; Asumadu-Sarkodie, S. A review of renewable energy sources, sustainability issues and climate change mitigation. *Cogent Eng.* **2016**, *3*, 1167990. [CrossRef]
32. Mendoza-Vizcaino, J.; Sumper, A.; Sudria-Andreu, A.; Ramirez, J.M. Renewable technologies for generation systems in islands and their application to Cozumel Island, Mexico. *Renew. Sustain. Energy Rev.* **2016**, *64*, 348–361. [CrossRef]
33. Child, M.; Haukkala, T.; Breyer, C. The role of solar photovoltaics and energy storage solutions in a 100% renewable energy system for Finland in 2050. In Proceedings of the 31st European Photovoltaic Solar Energy Conference and Exhibition, Hamburg, Germany, 14–18 September 2015. Available online: https://www.researchgate.net/publication/281859358_The_Role_of_Solar_Photovoltaics_and_Energy_Solutions_in_a_100_Renewable_Energy_System_for_Finland_in_2050 (accessed on 28 March 2021).
34. Behabtu, H.A.; Messagie, M.; Coosemans, T.; Berecibar, M.; Fante, K.A.; Kebede, A.A.; Mierlo, J.V. A Review of Energy Storage Technologies' Application Potentials in Renewable Energy Sources Grid Integration. *Sustainability* **2020**, *12*, 10511. Available online: https://www.mdpi.com/2071-1050/12/24/10511 (accessed on 28 March 2021). [CrossRef]
35. Makhijani, S.; Ochs, A.; Weber, M.; Konold, M.; Lucky, M.; Ahmed, A. *Jamaica Sustainable Energy Roadmap: Pathways to an Affordable, Reliable, Low-Emission Electricity System*; Worldwatch Institute: Washington, DC, USA, 2013. Available online: https://www.researchgate.net/publication/303811205_Jamaica_Sustainable_Energy_Roadmap_Pathways_to_an_Afordable_Reliable_Low-Emission_Electricity_System (accessed on 28 March 2021).
36. Hidalgo, D.B. Energía y desarrollo sostenible en Cuba. *Centro Azúcar* **2015**, *42*, 14–25. Available online: http://scielo.sld.cu/scielo.php?script=sci_arttext&pid=S2223-48612015000400002 (accessed on 30 March 2021).
37. Suárez-Rodríguez, A.J.; Beaton-Soler, P.A.; Faxas-Escalona, R. ESTADO Y PERSPECTIVAS DE LA ENERGÍA FÓSIL EN CUBA. *Tecnol. Quím.* **2011**, *XXXI*, 88–94.
38. Suárez, J.; Martín, G.J.; Sotolongo, J.A.; Rodríguez, E.; Savran, V.; Cepero, L.; Funes-Monzote, F.; Rivero, J.L.; Blanco, D.; Machado, R.; et al. Experiencias Del Proyecto BIOMAS-CUBA. Alternativas Energéticas a Partir de La Biomasa En El Medio Rural Cubano Experiences of the BIOMAS-CUBA Project. Energy Alternatives from Biomass in Cuban Rural Areas. *Pastos Forrajes* **2011**, *34*, 473–496, ISSN 0864-0394. Available online: http://scielo.sld.cu/scielo.php?pid=S0864-03942011000400007&script=sci_arttext&tlng=pt (accessed on 10 February 2021).

energies

Article

Study and Simulation of a Wind Hydro Isolated Microgrid

Rafael Sebastián and Antonio Nevado *

Departamento de Ingeniería Eléctrica, Electrónica y de Control, ETSII UNED, 28040 Madrid, Spain; rsebastian@ieec.uned.es

* Correspondence: anevado@ieec.uned.es

Received: 6 October 2020 ; Accepted: 11 November 2020; Published: 13 November 2020

Abstract: Isolated microgrids are microgrids which operate autonomously. This paper presents an isolated microgrid which combines a Hydraulic Turbine Generator (HTG) with a Wind Turbine Generator (WTG) to supply consumers forming a Wind Hydro Isolated Microgrid (WHIM). The WHIM includes a Dump Load (DL) to dissipate the active power excess. The WHIM has been modeled and its operation has been simulated in two modes: Wind-Hydro (WH), where both HTG and WTG supply power, and Wind-Only (WO) mode, where the WTG is the active power supplier and the HTG keeps connected to the grid with null power to generate the grid voltage. In WO, a fast frequency regulation is achieved by means of a controller which commands the DL to consume the WTG power excess. Additionally, the simulation of the mode transition from WO to WH, which is triggered by a system active power deficit in WO mode, is shown. A kick starting system designed to speed up the HTG power production improves the transient from WO to WH mode change. Finally, the simulations in WH mode show the interaction between the HTG and WTG. The two controls proposed have been proved effective and the simulations show a good WHIM dynamic performance.

Keywords: hydro turbine generator; wind turbine generator; dump load; isolated microgrids; power systems simulation; power systems control

1. Introduction

Isolated microgrids are microgrids which operate autonomously and are located in remote places [1–3]. Remote isolated microgrids have been mostly based on Diesel Generators (DG), which can be combined with sources of renewable energy, such as photovoltaic or wind [4] and short-term energy storage systems (ESS), mainly based on flywheels [5] or batteries [6,7]. When an isolated microgrid includes renewables and does not include a DG, it has to include another source of controlled generation, such as a fuel cell [8]. Among the renewable energies, hydro power is the only one that can produce fully controlled power. Hydro power is site-dependent since a river and the possibility of building a dam are needed. Wind power is also site-dependent, since the wind resource is needed at the considered place and, depending on the type of WTG used, can go from uncontrolled to partly controlled generation.

Figure 1 shows the isolated microgrid modeled and simulated in this paper, which combines hydro and wind power. It consists of a Hydraulic Turbine Generator (HTG), a Wind Turbine Generator (WTG), consumer load and a Dump Load (DL). The HTG comprises a Hydraulic Turbine (HT) which drives a Synchronous Machine (SM). The constant speed type WTG of Figure 1 comprises a fixed pitch Wind Turbine (WT) which drives an Induction Generator (IG). Both the HTG and the WTG are combined to form a Wind Hydro Isolated Microgrid (WHIM) to supply the isolated consumers. The DL comprises a resistor bank and a set of power switches and consumes active power when an excess of system active power exists.

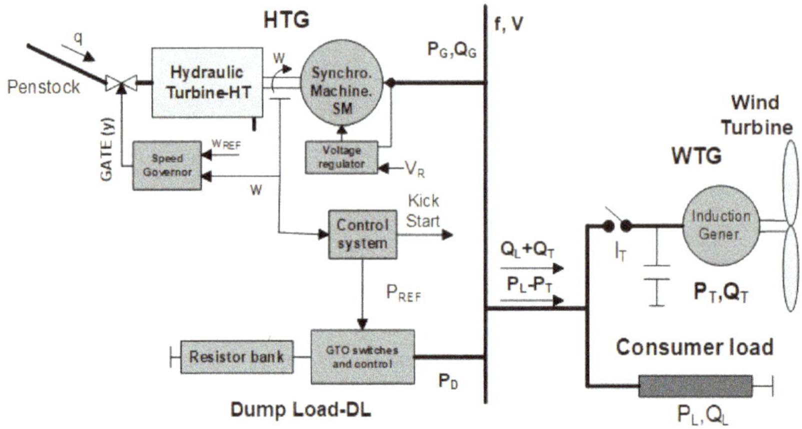

Figure 1. Wind-hydro power system scheme.

In the WHIM of Figure 1, the HTG is always running and connected to the isolated grid, since the SM creates the grid voltage waveform and the SM voltage regulator performs the system voltage regulation. Three operation modes are possible for the WHIM, shown in the figure: Hydro Only (HO), Wind Hydro (WH) and Wind Only (WO). In the HO mode, the HTG supplies all the consumer's demanded active and reactive power and the WTG is disconnected (I_T = OFF in Figure 1). In this mode, the WHIM behaves as an isolated hydropower system. Frequency regulation is performed by means of the HT speed governor, which actuates on the flow rate entering the HT, by means of a valve, to control the HT mechanical power produced. When the wind speed is above the WT cut-in speed, the WTG is able to supply power, so it is connected to the isolated grid changing to the WH mode. In the WH mode, the WTG supplies active power according to the existing wind speed and the HTG controls its active power to cover the active power net demand. Frequency regulation is achieved as in HO mode. In WH mode, the WTG produced power, P_T, can exceed that consumed by the load, P_L, ($P_T > P_L$) and if this happens, the DL must consume the WTG active power excess $P_T - P_L$ to guarantee the power system stability. If the condition $P_T > P_L$ persists, the WHIM is changed to WO mode. In WO mode, the WTG supplies the active power and the HTG keeps running with null flow rate and therefore null active power, in order to generate the grid voltage and supply reactive power. Frequency regulation is performed in this case by making the DL consume the WTG active power excess. If the WTG generated power falls below the power consumed by the load, then the HTG, that is already connected to the microgrid, must supply active power and the WHPS must change to WH mode.

HO mode is equivalent to an isolated Hydro Power System (HPS) and many small HPS operate isolated [9]. In some small isolated HPS [10] the HT has no flow regulation, so there is no active power regulation and the HTG works permanently at full power. In these cases, frequency regulation is performed by controlling the power consumed by the DL, so that the instantaneous sum of the power absorbed by the consumer load and the power dissipated in the DL is equal to the power generated by the HTG. A simulation example of this DL use in an isolated HPS can be found in [11]. This DL use is analogous to the use of the DL in the WO mode presented in this article, but with the difference that the WTG type used in this article produces non-controllable active power. In WH mode, the HTG counteracts the load and WTG power variations. In a previous paper published by one of the authors [12], a WHIM is modeled, but no DL is considered in the simulations. In that reference, WHIM transients in response to load increase in HO and WH modes are compared, and a better behavior in WH mode, due to the WTG damping action, is found. The WHIM model and its associated controls in this article are focused to allow operation in WO mode and to transition from WO to WH modes. WO mode simulations have been considered in several papers, both with systems which have a backup generator, such as a DG [13,14], or with systems where the WTG is the only power source [15,16]. When an isolated wind power system has no backup generator, it is mandatory for good system performance to include some energy storage: reference [15] includes a flywheel ESS whereas [16] includes a battery plus a supercapacitor ESS.

WHIM can be combined with other conventional power sources, as it is done at the wind-hydro-diesel isolated power system of el Hierro Island in Spain, which also includes a hydropower pumped-storage. The diesel off mode of El Hierro power system has been simulated in several papers, which mainly study the wind-hydro-pumped storage combination. Reference [17] studies the WO mode and shows the use of fixed and variable speed pumps integrated within the hydropower pumped-storage to regulate the system frequency. Reference [18] also studies the WO mode and a flywheel ESS is included in the simulations and different alternative frequency control schemes are studied and compared. References [17,18] show, among others, graphs of system frequency and active powers of the WTGs and pumps, but none of system voltage waveforms. The WHIM considered in this article has neither pump storage nor short-term energy storage and uses the DL as the variable controlled load to absorb the WTG power excess, being the lowest cost solution of all. Additionally, graphs of system voltage will be shown in the simulations section as high order electrical models for the electrical machines are used. In [19], logistic simulations of the El Hierro wind-hydro-diesel power plant in several operational modes are carried out in order to calculate the efficiency and the percentage of load demand covered by renewables in the different operational modes considered.

This article's main contribution is to present comprehensive simulations of a WHIM. These simulations firstly cover WO mode operation. In it, frequency regulation is achieved by means of the DL while the HTG actuates providing the grid voltage and as a backup generator. Secondly, the transition from WO mode to WH mode, which is triggered by the lack of system active power in WO mode, is simulated. Finally, WH mode is also simulated, and the interaction between the HTG and WTG is shown. Additionally, this article contributes with a proposal of a fast DL frequency regulator, which is active in WO mode, a control logic to start the WO to WH transition, and a control system aimed to speed up and smooth the transition from WO mode to WH mode.

Following this introductory section, this article is organized as follows: Section 2 presents the model of the HTG, along with a discussion on the most appropriate HT type for the WH and WO modes of operation and the models for the WTG and DL; Section 3 presents *Simulink* schematics for the simulated WHIM, along with the controls needed to operate in WO mode and to smooth the WO mode to WH mode transition; Section 4 presents the different simulation cases considered and mentioned above and, finally, Section 5 concludes with the main contributions of the paper.

2. WHIM Modeling

2.1. The HTG Model

A hydraulic turbine (HT) converts energy from falling water into rotating shaft power [10]. The HT model used in this article is nonlinear, since the simulations consider large variations for the HT operating point. The equations below, in per unit (p.u) values, describe the HT nonlinear model. Their derivation can be obtained from the references [12,20]:

$$q = y\sqrt{h} \quad (1)$$

$$P_{h-mec} = A_t\eta q h - K_D y(\omega - 1) \quad (2)$$

Equation (1) describes the valve which regulates the flow rate q to the HT. In it, y is the turbine gate opening position that can vary from 0 (fully closed) to 1 (fully open); $q = Q/Q_{base}$, is the flow rate p.u., Q being the flow rate passing through the turbine and Q_{base} the turbine flow rate with the gate fully open, $h = H/H_{base}$ is the pressure head of water p.u., with H as the pressure head in the turbine admission and H_{base} the total available static head above the turbine [20]. Equation (2) gives the mechanical power p.u. produced by the HT, P_{h-mec}, which has two terms. The first term is the power produced by the flow rate p.u q, with an effective pressure head of water in p.u. h and η is the turbine hydraulic efficiency. A_t is a proportionality factor which depends on the HT active rated power and the SM apparent rated power. The second term accounts for a speed deviation effect $(\omega - 1)$ where ω is the HT shaft speed p.u; 1 p.u. is the rated turbine speed and K_D, the damping torque coefficient [20].

The penstock is the pressure pipe that delivers water from the dam to the turbine admission. The following equation, describing the penstock model, was previously derived in [12,20]:

$$\frac{dq}{dt} = \frac{1}{T_w}\left(1 - h - h_f\right) \quad (3)$$

In it, h_f is the head loss due to friction in the penstock and T_w is the water time constant or water starting time in seconds, defined as:

$$T_w = \frac{L}{Ag}\frac{Q_{base}}{H_{base}} \quad (4)$$

where L, in m and A, in m^2, are the penstock length and area, respectively, and g is the gravitational acceleration. Equation (3) is valid for penstocks of short length, in which the pressure wave effects can be neglected.

In WH mode, the HTG must supply the grid with an amount of power equal to the difference between the consumer load and the WTG supplied power. The HT working power range can therefore be very broad and the low load regime can be very common. The efficiency of a HT depends mainly on the flow rate [10]. The HT efficiency is null for flow rates ranging from zero to the no-load flow rate q_{nl}. Different types of hydro-turbines, such as Pelton, Crossflow, Francis, Kaplan, etc. have different q_{nl} and efficiency curves. Due to the highly variable working regime of the HT in WH mode, a good efficiency for $q > q_{nl}$ is the main requirement for the HT in the study considered here. Additionally, in WO mode, as commented before, the HT runs with zero flow rate. When the WHIM is in WO mode and the WTG active power falls below the consumer load, the system frequency will fall and the WHIM must transition from WO to WH mode, ordering the HTG to supply power. In order to supply power, the HT speed governor must increase the flow rate from zero. The time spent to take the flow rate from zero to q_{nl} is a dead time in which the HT is not producing any power and the system frequency will continue falling. The bigger q_{nl}, the longer the dead time, and the worse will be the system transient. Therefore, an HT with a small q_{nl} is needed to improve the WO to WH modes transition. Among the different HT types, the ones that most comply with the previous requirements are Crossflow and Pelton impulse type turbines, Pelton type HT having better efficiency figures than Crossflow type HT. A Pelton type HT has, therefore, been considered in this article. Figure 2 depicts the Pelton turbine efficiency curve used in this article, as a function of q, in abscissa, where q_{nl} is 0.1 p.u.

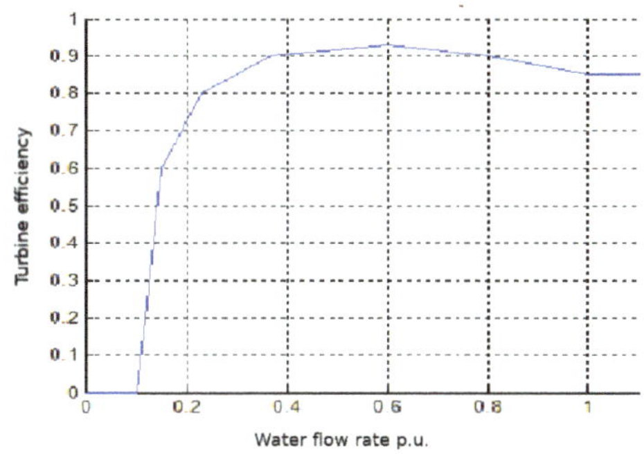

Figure 2. Pelton turbine efficiency vs. water flow rate p.u.

The HT mechanical output power is converted into electrical power by the SM. The SM must be permanently connected to the isolated grid as it generates the isolated grid voltage waveform. To allow the SM automatic voltage regulator control the system voltage within allowable limits, the SM speed has to always be close to its rated speed. The SM nominal power P_{SM-NOM} is 300 kVA, which is enough to provide the microgrid with all the active and reactive power demanded. Since the system frequency is $f = 50$ Hz and the SM has $p = 16$ pole pairs, calculating the synchronous speed, n, as $n = 60f/p$, the HTG rated speed, n_{ref}, results in 187.5 rpm. Inertia constant H for low speed HTGs (<200 rpm), ranges between 2–3 s [21], so 2 s is assigned to H_{HTG}, as this article's HTG is a low power one.

2.2. The WTG Model

The WTG in Figure 1 is a fixed-pitch WT driving an Induction Generator (IG) directly connected to the autonomous grid. Both elements, WT and IG, form a constant-speed stall-controlled WTG. The mechanical power extracted from the wind by the wind turbine, as stated by [22], is

$$P_{t-mec} = C_p\, P_{wind} = C_p\, \frac{1}{2}\rho A v^3 \tag{5}$$

where the power coefficient, C_p, represents the ratio of the power extracted from the wind to the power available in the wind. The power available in the wind, P_{wind}, can be calculated as half of the air density ρ, times the blade swept area A, times the cube of the wind speed v. In the absence of pitch regulation and due to the very limited IG speed range variation, C_p can be considered, as a first approximation, a function of the wind speed only. As the wind speed is quasi-random, the active power generated by the WTG will be uncontrolled. The stall-controlled WT model simulated here follows the description presented in [23]. The IG has a rated power of 275 kW (WTG rated power P_{T-NOM} = 275 kW), which is enough to supply all the active power demanded in WO mode, and since it consumes reactive power, a capacitor bank is included in Figure 1 to improve the power factor. Inertia constant H_w values for WTGs range from 2 to 6 s [24], so 2 s is assigned to H_w, as this article's WTG is a low power one.

2.3. The DL Model

The Dump Load (DL) is used in the WHIM to artificially load the isolated grid when the WTG power exceeds the load consumed power in WO mode. The DL shown in Figure 1 comprises a bank of eight three-phase resistors connected in series with Gate Turn-Off (GTO) thyristor-based power switches. The eight resistor values are $R, R/2, R/2^2, ..., R/2^7$. With V_n the isolated grid rated voltage, $P_0 = V_n^2/R$

is the rated power of the resistor of value R and the eight resistors rated powers are $P_0, 2P_0, 2^2P_0, ..., 2^7P_0$. If $I_0, I_1, ..., I_7$ are the opened(0)/closed(1) states of the GTO power switches in series with the resistors, the active power absorbed by the DL can be expressed as

$$(I_0 + I_1 2^1 + ... + I_7 2^7)P_0 \tag{6}$$

Equation (6) states a DL power variation from 0 to $255P_0$. In this article $P_0 = 1.5$ kW and the DL rated power P_{D-NOM} is 382.5 kW, which is 39% greater than P_{T-NOM}. This P_{D-NOM} value guarantees the possibility of WO mode operation, even in the absence of any consumer load and with a WTG power rated above one.

3. Simulation Schematics

The isolated WHIM shown in Figure 1 was simulated using the *MATLAB-Simulink* framework [25]. All the parameters of the presented WHIM are shown in Appendix A. The WHIM Simulink schematic is shown in Figure 3. Some of the components described above, such as the 275 kW IG of the WTG, the 300 kVA SM and its voltage regulator, the consumer load, etc. are blocks which belong to the *SimPowerSystems* blockset for Simulink. The electrical part of the IG is represented by a fourth-order model, the SM electrical part is represented by a sixth-order model and its voltage regulator-exciter is an IEEE type 1 model.

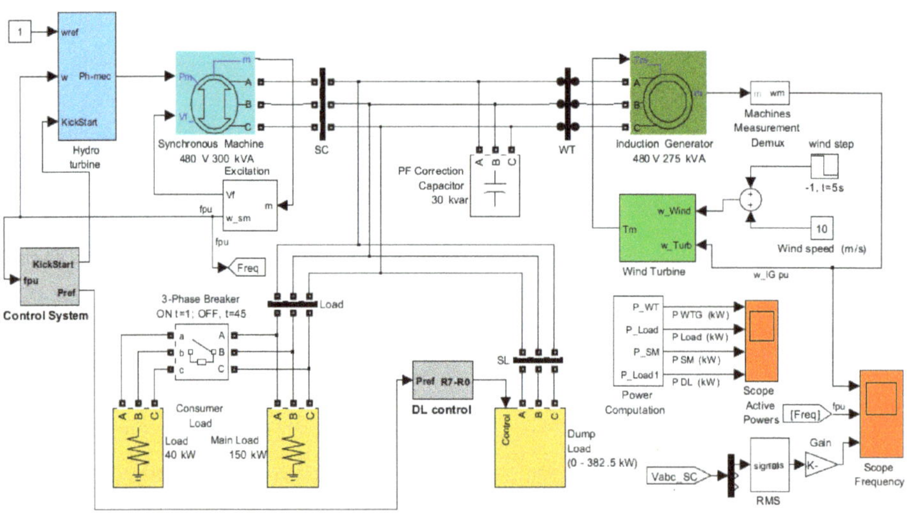

Figure 3. Wind-hydro isolated microgrid *Simulink* schematics.

The Hydro Turbine block implements the HT, the gate, the penstock and the speed governor models. It receives as inputs the constant 1 p.u. speed reference and the current HTG speed ω and produces as an output the mechanical power P_{h-mec} needed to bring the HTG speed to its set point. Figure 4 shows the schematics of the non-linear hydraulic turbine model, which includes Equations (1) and (2) and penstock model Equation (3).

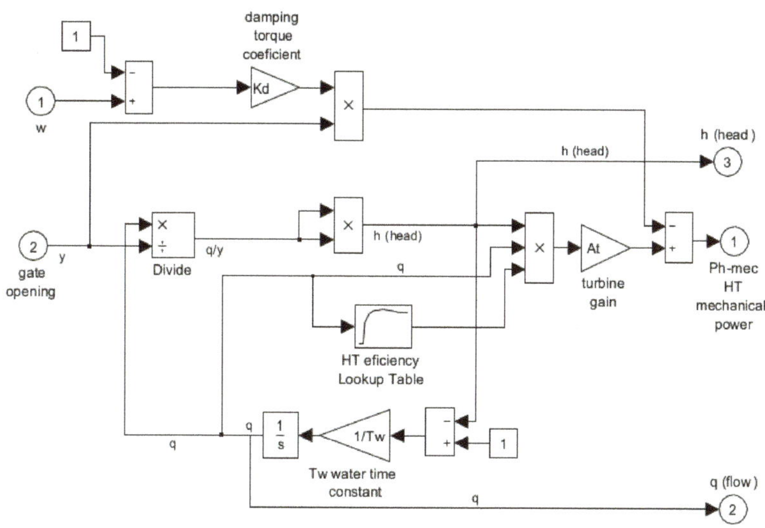

Figure 4. Hydraulic turbine *Simulink* schematics.

In the simulations below, the penstock friction losses term h_f in Equation (3) and the damping torque coefficient K_D in Equation (2) are considered negligible. As an effect, since these natural dampings are not implemented, the system stability of the simulated WHIM can be considered worse than in reality. The Pelton turbine efficiency has been implemented by means of a lookup table, which follows the efficiency curve depicted in Figure 2.

Figure 5 shows the HT speed governor schematic, which comprises a Proportional Integral Derivative (PID) speed controller and a servo actuator. The PID speed controller performs an isochronous speed control, so that in steady state, the HT speed will be the rated one, assuming the HT demanded power is in the range $[0, HT_{NOM}]$, being HT_{NOM} the HT rated power. The PID pure derivative term has been replaced by a derivative filter. PID parameters K_p, K_i and K_d, are calculated as proposed in [20]:

$$K_p = 1.6\frac{H}{T_w} \quad (7)$$

$$K_i = 0.14\frac{K_p}{T_w} \quad (8)$$

$$K_d = 0.54H \quad (9)$$

where $H = 2$ s is the HTG inertia constant previously justified and T_w is the previously defined water time constant. The penstock is assumed to be short, with $T_w = 1$ s. With these values, Equations (7)–(9) are used to calculate the PID speed controller parameters. The PID input is the sum of the HTG speed error and the *Kick Start* (KS) signal. As shown in Figure 5, the KS signal is produced by a second order system with a zero at the origin, which acts as a derivative. When the WHIM transitions from WO to WH mode, the WH/WO* mode signal changes from 0 to 1 and the KS signal waveform is the impulse response of the second order system formed only by the denominator. As stated above, just before the WO to WH transition occurs, the flow rate is null and the aim of the KS signal is to speed up the increase in flow rate from zero to q_{nl}, reducing the dead time in which there is no production of mechanical power. In the simulations below, the benefits of using the KS signal are shown. In the figure, it can be seen how the servo converts the PID output into gate opening y, constrained to the range $[0, 1]$. The servo-motor speed limits are included to avoid in the penstock big pressure transients and water hammer [26]. This servo-motor speed limits are a strong constraint to any speed controller to be implemented here.

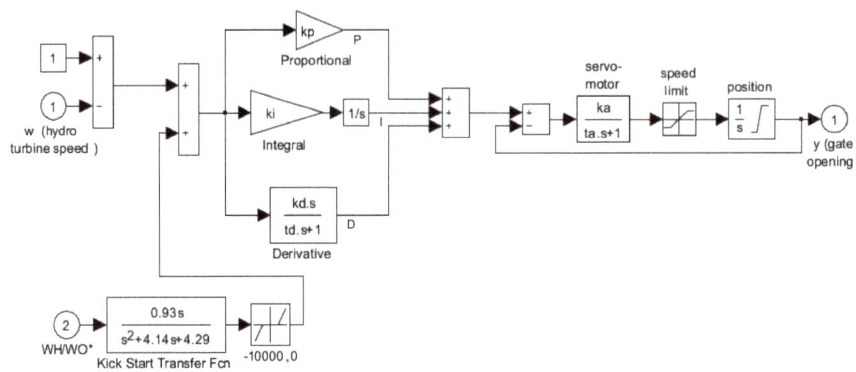

Figure 5. Hydraulic turbine speed governor *Simulink* schematics.

The WT block of Figure 3 is implemented with the WT power curves of Figure 6, which define the WT shaft mechanical power P_{t-mec} as a function of the wind speed and the wind turbine rotational speed. To calculate the torque T_m delivered to the WTG-IG, P_{t-mec} is divided by the WT speed.

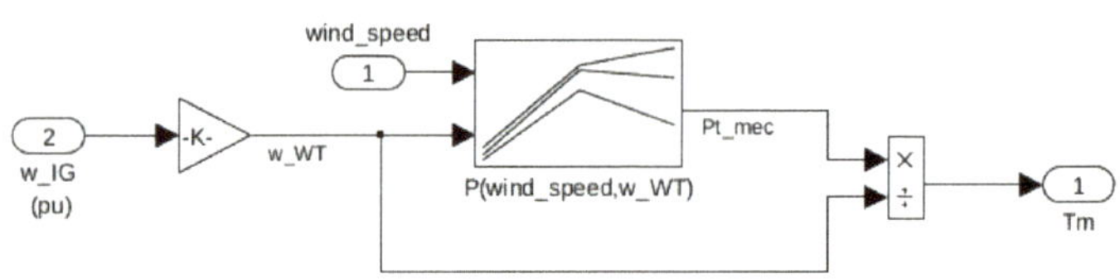

Figure 6. Wind turbine mechanical torque calculation.

DL control in WO mode is performed by the PID depicted in Figure 7 with the objective of regulating the system frequency. The rated frequency 1 p.u. is subtracted from the system frequency in order to obtain the frequency error, which is the input to the DL PID. The DL PID outputs, when positive, the reference power P_{ref} to be consumed by the DL. The proportional, derivative and integral constants of the DL PID have been tuned to speed up the frequency response, so it is much faster than the HT PID one, and to minimize frequency overshoots. A negative output produced by the DL PID implies that active power must be produced instead of consumed, to regulate the system frequency and therefore indicates active power deficit. Since the DL cannot produce any power, a DL PID negative output means that the HTG must produce power in order to achieve the proper active power balance, and therefore a transition from WO to WH mode is needed. An active power deficit situation leads to a frequency fall, which is detected by comparing the output of the integral component of the DL PID which integrates the frequency error with the constant -0.1, as shown in Figure 7. An active power deficit signal is activated when the output of the integrator is below the constant −0.1, which makes the mode flip-flop output toggle from 0 to 1, making in turn the system change to WH mode. The constant −0.1 has been adjusted taken into account the system dynamics in WO mode. In WH mode, the power reference for the DL is zero.

Finally it is worth noting that the HT PID is active in the three modes of operation. However, in WO mode, the DL PID controls the system frequency in a much faster way than the HT PID does, so the HT PID does not have any impact during WO mode.

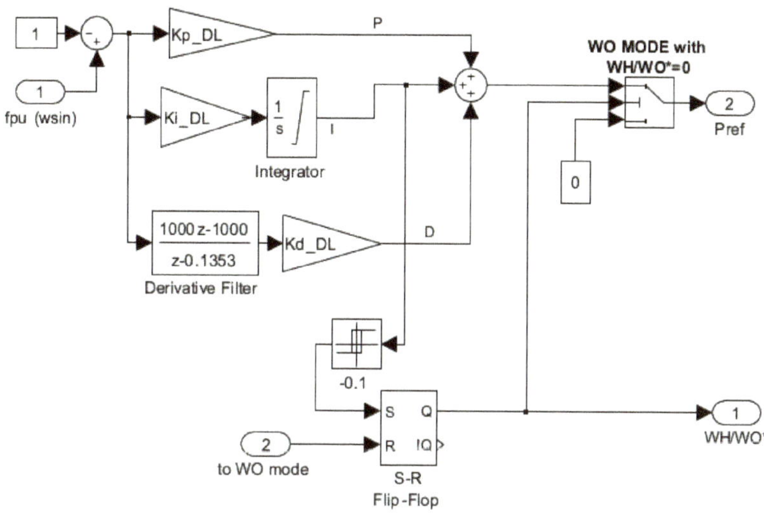

Figure 7. Dump load control *Simulink* schematics.

4. Simulation Results

Simulation results are shown in Figures 8–10, considering the following variables: frequency per unit (fpu) in Figure 8, Root Mean Square (RMS) voltage p.u. in Figure 9 and active powers in kW for the WTG, HTG, DL and consumer load in Figure 10. In Figure 10, generated and consumed active powers are plotted positive and negative, respectively, so that in steady state the active powers sum is null. At the test starting point, the WHIM is in WO mode (flip-flop mode output 0 in Figure 7), so the HTG active power and flow rate are null, the consumer load and the DL are consuming 150 kW and 50 kW, respectively, and the WTG is generating an active power of 200 kW with a wind speed of 10 m/s. The system is in steady state.

In the following simulation results, the settling time has been chosen to be the time it takes for the system frequency to fall within a neighborhood of 0.01% around 1 p.u. steady state value after the event that produced the transient under discussion.

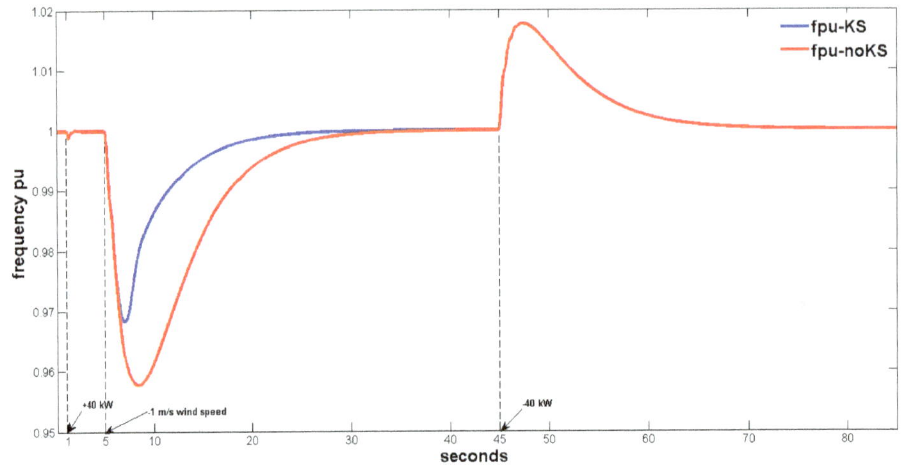

Figure 8. Microgrid frequency in p.u.

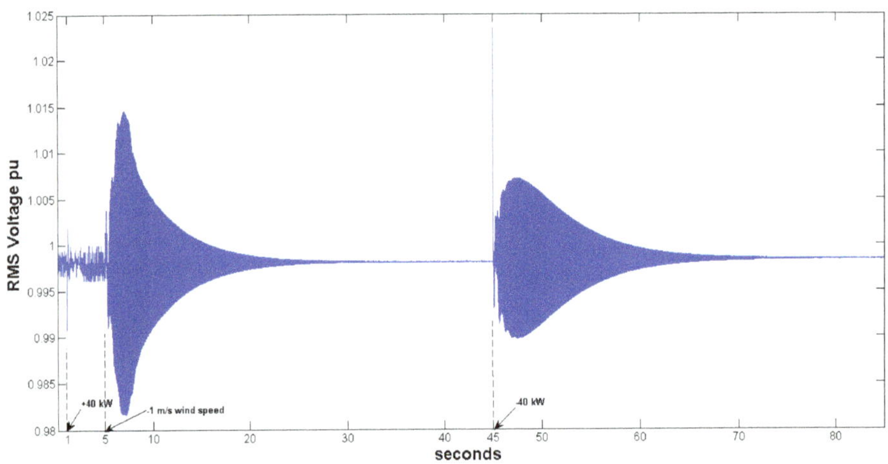

Figure 9. Microgrid RMS voltage in p.u.

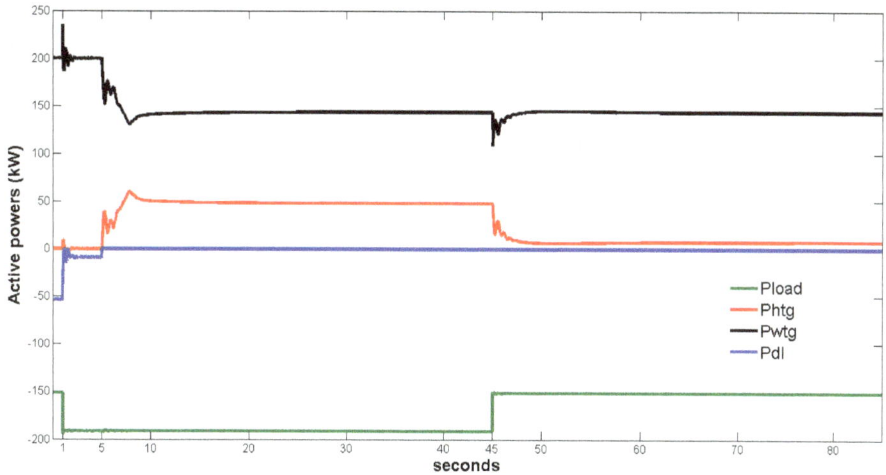

Figure 10. Active powers in WTG, HTG, consumer load and dump load in kW.

4.1. Simulations in WO Mode

The three-phase breaker in Figure 3 is closed at $t = 1$ s, so that an extra 40 kW resistive load is connected to the system, as it can be observed in the load active power curve in Figure 10. The DL controller reacts by reducing the DL consumed power in the same quantity that the consumer load has increased, being the DL active power final value of 10 kW in steady state. Figure 10 also shows a positive peak transient in the WTG active power after the 40 kW positive load step, which counteracts the active power deficit and improves the frequency response. This effect is due to the IG present in the WTG, which provides a damped response [27]. The WTG power in steady state is 200 kW, as initially, since the wind speed does not change. In the transient, the fpu shows a minimum of 0.9988, the voltage maximum and minimum values are 1.0018 and 0.9908, respectively, and the settling time in this case is 0.975 s. The active power deficit detector does not activate.

4.2. WO to WH Mode Transition

At $t = 5$ s, wind speed decreases from 10 to 9 m/s and, consequently, the WTG active power decreases too, and gets smaller than the consumer load. The system frequency falls due to the active power deficit, so that DL PID control reduces the DL consumed power until it reaches zero. The system frequency fall activates the active power deficit detector, so the WHIM mode flip-flop output changes to 1, starting the WO to WH mode transition. The positive edge of the WH/WO* signal triggers the HT KS system, whose output is added, as stated above, to the frequency error to be the HT PID input. Figure 10 shows a HTG power increasing with some initial oscillations during the transient. At steady state, which is reached 26.488 s after the negative wind step, the HTG final power is 47 kW. The final WTG active power is 143 kW. Due to voltage oscillations, the load power shows small variations too during the transient, but eventually settles at the initial value of 190 kW. The system frequency presents a minimum of 0.968 p.u. and the minimum and maximum voltages obtained are 0.9817 and 1.0146, respectively. Figure 8 also shows the system frequency with the KS function deactivated in red line, in which the frequency error is the only variable considered by the HT-PID. In the no-KS case, the minimum frequency is 0.9577, the minimum and maximum voltages are 0.9758 and 1.0195 and the settling time is 32.844 s. The frequency transient in the noKS case is therefore 24% longer, and the voltage and frequency variations are bigger, which justifies the necessity of using the KS system for the WO mode to WH mode transition.

As can be seen, the transient duration occurring in WO mode is more than 26 times shorter than the one occurring in the transition from WO mode to WH mode. This fact indicates that both controllers, HT PID and DL PID, operate in different time scales and therefore supports the decision stated in Section 3 of keeping both active in WO mode.

4.3. The WH Mode Simulation

At t = 45 s the circuit breaker in Figure 3 is opened, thus reducing the consumer load in 40 kW. The WTG active power reacts firstly with a negative peak, which temporarily compensates the consumer load reduction. Again in this case, the IG provides damping. In steady state, this reached 30.89 s after the load reduction and the WTG remained at the initial value of 143 kW, since the wind speed does not change. Figure 10 shows a HTG power decreasing with oscillations at the beginning of the negative load step, and finally, the HTG assumes the load reduction with a final power of 7 kW. The fpu maximum is 1.0177 and the RMS voltage p.u. minimum and maximum values are 0.9898 and 1.0235, respectively.

Table 1 summarizes the results obtained with the simulations.

Table 1. Simulation results summary.

Variable \Case	WO	WO2WH KS	WO2WH no-KS	WH
Δf	[0.9988, 1]	[0.9683, 1]	[0.9577, 1]	[1, 1.0177]
Δf [%]	−0.12	−3.17	−4.23	+1.77
ΔV	[0.9908, 1.0018]	[0.9817, 1.0146]	[0.9758, 1.0195]	[0.9898, 1.0235]
ΔV [%]	1.1	3.29	4.37	3.37
t_s [s]	0.975	26.487	32.8440	30.892

4.4. Hydraulic Variables

The flow rate and HT mechanical power plots for the KS and no-KS cases in solid and dotted line, respectively, are presented until t = 45 s in Figure 11. As commented on, the frequency transient after the connection of +40 kW at t = 1 s is managed by the DL PID, but the small frequency variations also causes the HT PID to slightly react and this explains the small perturbation in the flow rate after t = 1, as seen in Figure 11. At t = 5.1 s both KS and no-KS flow rate curves left the zero value due to the response of the HT-PID to the frequency falling after the −1 m/s wind step at t = 5 s. The flow rate in the KS case increases faster due to the added signal from the KS transfer function triggered by the active power deficit detector activation. The time when the flow rate reaches the q_{nl} value (0.1 p.u.) is for the KS and no-KS cases 6.2 and 6.43 s, respectively. From that time on, the mechanical power produced by the HT, P_{h-mec}, rises from the initial value of zero as can be also seen in Figure 11, so the interval from t = 5 s until times 6.2/6.43 s is a dead time where there is a system active power deficit, as the WTG power does not reach the demanded active power and therefore the system frequency falls due to the imbalance. Now, it is seen by means of graphics the importance of a small q_{nl} in the HT efficiency curve commented in Section 2.1 in order to minimize the negative frequency peak in the WO to WH mode transition. In the HT mechanical power plots, it is also seen that the KS case follows the flow rate curve with less delay and the reason is a better transient of the head at the turbine admission.

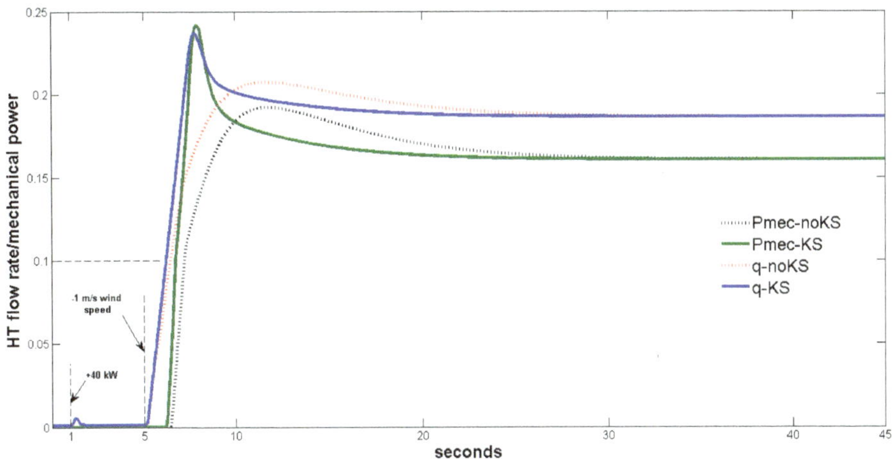

Figure 11. Hydraulic variables.

5. Conclusions

In this article, the modeling of a WHIM has been presented, with detailed models for the HT, the penstock, the WT and DL, along with its associated controls. The Pelton type HT was chosen for its small q_{nl} and good efficiency. The WHIM has been simulated using the MATLAB-Simulink environment.

In WO mode, both the DL-PID and the HT-PID work but because the DL-PID is around 24 times faster than the HT-PID, the final result is that the frequency regulation is achieved by the DL and its associated PID.

In WO mode to WH mode transition, the simulation has shown that the KS/no-KS frequency minimum values are 0.968/0.958 and the setup times are 26.487/32.844, so the KS case results in a much better transient. Therefore, the KS system has a fundamental role in the WO to WH transition. It is also seen in Section 4.4 the importance of using a HT with a small q_{nl} in order to speed up the transient. In WH mode, the simulation shows that the isochronous HT-PID regulates the frequency while the HTG accommodates its power to the net consumed load.

Author Contributions: Conceptualization, R.S.; methodology, R.S.; software, A.N.; validation, R.S. and A.N.; formal analysis, R.S.; investigation, R.S. and A.N.; resources, R.S.; writing—original draft preparation, R.S.; writing—review and editing, A.N.; visualization, R.S.; supervision, A.N. All authors have read and agreed to the published version of the manuscript.

Funding: This research received no external funding.

Conflicts of Interest: The authors declare no conflict of interest.

Appendix A. System Configuration

Appendix A.1. Isolated Microgrid

Rated frequency, f_{NOM} = 50 Hz.

Rated voltage (RMS, phase to phase) = 480 V.

Appendix A.2. Hydro Turbine Generator

Water time constant T_w = 1 s.
Proportionality factor (turbine gain), A_t =1.375.
HTG Inertia constant, H = 2 s.
HTG rated speed = 187.5 rpm.
SM rated power, P_{SM-NOM} = 300 kVA.
SM number of pole pairs = 16.

Appendix A.3. Wind Turbine Generator

WTG rated power, P_{T-NOM} = 275 kW.
WTG Inertia constant, H_w = 2 s.

Appendix A.4. Dump Load

DL rated power = 357 kW.
DL-PID parameters K_p = 446; K_i = 1990; K_d = 15.

References

1. Lasseter, R. Microgrids. In *Proceedings of the IEEE PES Winter Meeting*; IEEE: New York, NY, USA, 2002; Volume 1, pp. 305–308.
2. Basak, P.; Chowdhury, S.; Halder nee Dey, S.; Chowdhury, S.P. A Literature Review on Integration of Distributed Energy Resources in the Perspective of Control, Protection and Stability of Microgrid. *Renew. Sustain. Energy* **2012**, *16*, 5545–5556. [CrossRef]
3. Piagi, P.; Lasseter, R.H. Autonomous Control of Microgrids. In *PES Meeting, Montreal*; IEEE: New York, NY, USA, 2006.
4. Sebastián, R.; García-Loro, F. Review on Wind Diesel Systems Dynamic Simulation. In *IECON 2019—45th Annual Conference of the IEEE Industrial Electronics Society*; IEEE: New York, NY, USA, 2019; pp. 2489–2494.
5. Sebastián, R.; Peña-Alzola, R. Flywheel Energy Storage and Dump Load to Control the Active Power Excess in a Wind Diesel Power System. *Energies* **2020**, *13*, 2029. [CrossRef]
6. Sebastián, R. Battery Energy Storage for Increasing Stability and Reliability of an Isolated Wind Diesel Power System. *IET Renew. Power Gener.* **2017**, *11*, 296–303. [CrossRef]
7. Bø, T.I.; Johansen, T.A. Battery Power Smoothing Control in a Marine Electric Power Plant Using Nonlinear Model Predictive Control. *IEEE Trans. Control. Syst. Technol.* **2016**, *25*, 1449–1456. [CrossRef]
8. Taghizadeh, M.; Mardaneh, M.; Sha Sadeghi, M. Frequency Control of a New Topology in Proton Exchange Membrane Fuel Cell/Wind Turbine/Photovoltaic/Ultra-Capacitor/Battery Energy Storage System Based Isolated Networks by a Novel Intelligent Controller. *J. Renew. Sustain. Energy* **2014**, *6*, 053121. [CrossRef]
9. Hasmaini, M.; Hazlie, M.; Bakar, A.H.; Ping, H.W. A Review on Islanding Operation and Control for Distribution Network Connected with Small Hydro Power Plant. *Renew. Sustain. Energy Rev.* **2011**, *15*, 3952–3962.
10. Paish, O. Small Hydro Power: Technology and Current Status. *Renew. Sustain. Energy Rev.* **2002**, *6*, 537–556. [CrossRef]
11. Doolla, S.; Bhatti, T.; Bansal, R. Load Frequency Control of an Isolated Small Hydro Power Plant Using Multi-pipe Scheme. *Electr. Power Components Syst.* **2011**, *39*, 46–63. [CrossRef]
12. Sebastián, R.; Quesada, J. Modeling and Simulation of an Isolated Wind Hydro Power System. In *IECON 2016-42nd Annual Conference of the IEEE Industrial Electronics Society*; IEEE: New York, NY, USA, 2016; pp. 4169–4174.
13. Sebastián, R. Application of a Battery Energy Storage for Frequency Regulation and Peak Shaving in a Wind Diesel Power System. *IET Gener. Transm. Distrib.* **2016**, *10*, 764–770. [CrossRef]

14. Lukasievicz, T.; Oliveira, R.; Torrico, C. A Control Approach and Supplementary Controllers for a Stand-Alone System with Predominance of Wind Generation. *Energies* **2018**, *11*, 411. [CrossRef]
15. Sebastián, R.; Peña-Alzola, R. Flywheel Energy Storage Systems: Review and Simulation for an Isolated Wind Power System. *Renew. Sustain. Energy Rev.* **2012**, *16*, 6803–6813. [CrossRef]
16. Mendis, N.; Muttaqi, K.M.; Perera, S. Management of Battery-Supercapacitor Hybrid Energy Storage and Synchronous Condenser for Isolated Operation of PMSG Based Variable-Speed Wind Turbine Generating Systems. *IEEE Trans. Smart Grid* **2014**, *5*, 944–953. [CrossRef]
17. Sarasúa, J.; Martínez-Lucas, G.; Platero, C.; Sánchez-Fernández, J. Dual Frequency Regulation in Pumping Mode in a Wind–Hydro Isolated System. *Energies* **2018**, *11*, 2865. [CrossRef]
18. Sarasúa, J.; Martínez-Lucas, G.; Lafof, M. Analysis of Alternative Frequency Control Schemes for Increasing Renewable Energy Penetration in El Hierro Island Power System. *Int. J. Electr. Power Energy Syst.* **2019**, *113*, 807–823. [CrossRef]
19. Briongos, F.; Platero, C.A.; Sánchez-Fernández, J.A.; Nicolet, C. Evaluation of the Operating Efficiency of a Hybrid Wind–Hydro Powerplant. *Sustainability* **2020**, *12*, 668. [CrossRef]
20. Mover, W.G.P.; Supply, E. Hydraulic Turbine and Turbine Control Models for System Dynamic Studies. *IEEE Trans. Power Syst.* **1992**, *7*, 167–179.
21. Kothari, D.P.; Nagrath, I.J. *Modern Power System Analysis*; Tata McGraw-Hill Education: New York, NY, USA, 2003.
22. Rodríguez-Amenedo, J.L.; Burgos-Díaz, J.C.; Arnalte-Gómez, S. *Sistemas Eólicos de Producción de Energía Eléctrica*; Rueda: Madrid, Spain, 2003; ISBN 9788472071391.
23. Gagnon, R.; Saulnier, B.; Sybille, G.; Giroux, P. Modeling of a Generic High-Penetration No-Storage Wind-Diesel System Using Matlab Power System Blockset. In Proceedings of the 2002 Global Windpower Conference, Paris, France, 2–5 April 2002.
24. Knudsen, H.; Nielsen, J.N. Introduction to the Modeling of Wind Turbines. In *Wind Power in Power Systems*; Wiley: Chicester, UK, 2005; pp. 525–585.
25. MathWorks, Model-Based Simulation and Design. Available online: https://mathworks.com/products/simulink.html (accessed on 27 September 2020).
26. Platero, C.A.; Nicolet, C.; Sánchez, J.A.; Kawkabani, B. Increasing Wind Power Penetration in Autonomous Power Systems Through No-Flow Operation of Pelton Turbines. *Renew. Energy* **2014**, *68*, 515–523. [CrossRef]
27. Margaris, I.D.; Papathanassiou, S.A.; Hatziargyriou, N.D.; Hansen, A.D.; Sorensen, P. Frequency Control in Autonomous Power Systems with High Wind Power Penetration. *IEEE Trans. Sustain. Energy* **2012**, *3*, 189–199. [CrossRef]

energies

MDPI

Article

An Adaptive Control Scheme for Variable Speed Wind Turbines Providing Frequency Regulation in Isolated Power Systems with Thermal Generation

Ana Fernández-Guillamón [1,†], Guillermo Martínez-Lucas [2,*,†] and Ángel Molina-García [1,†] and Jose Ignacio Sarasua [2,†]

1 Department of Automatics, Electrical Engineering and Electronic Technology, Universidad Politécnica de Cartagena, 30202 Cartagena, Spain; ana.fernandez@upct.es (A.F.-G.); angel.molina@upct.es (Á.M.-G.)

2 Department of Hydraulic, Energy and Environmental Engineering, Universidad Politécnica de Madrid, 28040 Madrid, Spain; joseignacio.sarasua@upm.es

* Correspondence: guillermo.martinez@upm.es; Tel.: +34-91-067-43-33

† These authors contributed equally to this work.

Received: 27 May 2020; Accepted: 28 June 2020; Published: 1 July 2020

Abstract: The lack of synchronous inertia, associated with the relevant penetration of variable speed wind turbines (VSWTs) into isolated power systems, has increased their vulnerability to strong frequency deviations. In fact, the activation of load shedding schemes is a common practice when an incident occurs, i.e., the outage of a conventional unit. Under this framework, wind power plants should actively contribute to frequency stability and grid reliability. However, the contribution of VSWTs to frequency regulation involves several drawbacks related to their efficiency and equipment wear due to electrical power requirements, rotational speed changes, and subsequently, shaft torque oscillations. As a result, wind energy producers are not usually willing to offer such frequency regulation. In this paper, a new control technique is proposed to optimize the frequency response of wind power plants after a power imbalanced situation. The proposed frequency controller depends on different power system parameters through a linear regression to determine the contribution of wind power plants for each imbalance condition. As a consequence, VSWTs frequency contribution is estimated to minimize their mechanical and electrical efforts, thus reducing their equipment wear. A group of sixty supply-side and imbalance scenarios are simulated and analyzed. Results of the case study are compared to previous proposals. The proposed adaptive control reduces the maximum torque and rotational speed variations while at the same time maintaining similar values of the load shedding program. Extensive results and discussion are included in the paper.

Keywords: frequency control; isolated system; linear regression; power system stability; wind turbines

1. Introduction

The different services carried out by the transmission system operators (TSO) for a reliable and secure power system are known as ancillary services [1]. Among them, load-frequency control focuses on mitigating the effects of unpredictable changes both in the demand and in the generation units that can address frequency deviations [2]. In fact, power imbalances between generation and consumption cause frequency variations [3]. In Europe, frequency control has a hierarchical structure, usually organized in up to five layers (from fast to slow timescales): (i) frequency containment (also known as primary frequency control); (ii) imbalance netting; (iii) automatic and/or manual frequency restoration (also known as secondary frequency control), and (iv) replacement [4]. If the different reserves of such frequency control layers are consumed or unable to keep frequency within an acceptable range, a variety of strategies called special protection systems are then used. Load shedding is included in those special protection systems. Moreover, it is considered as the last option to prevent frequency instability [5]. Despite load shedding being an effective solution to prevent a power system collapse after a major imbalance, it is considered as an undesirable situation and it is important to reduce it as much as possible [6–8].

Traditionally, power systems have been based on conventional power plants with synchronous generators directly connected to the grid, automatically providing their stored kinetic energy after a generation-load mismatch [9]. However, in recent decades, power systems have been suffering a slow change from conventional synchronous power plants to inverter-interfaced renewable energy sources (II-RES), i.e., wind power plants based on variable speed wind turbines (VSWTs) and/or solar photovoltaic (PV) [10]. Among them, VSWTs are considered as the most efficient, developed, and installed renewable resource, and currently they account for more than 650 GW of installed capacity around the world [11,12]. This remarkable integration of wind power plants requires an important reformulation of their contribution to ancillary services [13]. Moreover, as they are connected to the grid through power inverters, the synchronous inertia of the power system decreases when such renewable source replaces conventional power plants [14]. Indeed, faster rate of change of frequency (RoCoF) and larger frequency deviations are related to low synchronous system inertia values [15]. These effects are even more critical in isolated power systems [16,17]. As a result, Toulabi et al. consider that, due to the massive integration of VSWTs, their participation into frequency control is necessary [18]. With this aim, different frequency control approaches can be found in the specific literature to effectively replace conventional power plants by VSWTs and maintain a reliable power system operation [19]. These strategies are summarized in Figure 1 according to the different approaches [20,21].

- VSWTs frequency control strategies
  - De-loading
    - Over-speed
    - Pitch angle
  - Inertia response
    - Droop
    - Hidden inertia emulation
    - Fast power reserve

Figure 1. General classification for Variable Speed Wind Turbines (VSWTs) frequency control techniques.

VSWTs are designed to work in their maximum power point (MPP) according to the available wind speed s_w: $p_{MPP}(s_w)$ [22]. As a consequence, the first approach (deloading technique) consists of operating the VSWTs in a suboptimal power point p_{del}, below $p_{MPP}(s_w)$. Therefore, a certain amount of power Δp_d can be supplied in case of a power imbalance [23,24]. Two different possibilities are identified [25]: (i) the pitch-angle control and (ii) over-speed control. In the first one, the pitch angle is increased from β_0 to β_1 for a constant s_w. Subsequently, the generated power p_{del} is below the maximum power p_{MPP} [26–29]. When the additional power Δp_d is supplied, the pitch angle reduces to β_0. The over-speed control increases the rotational speed of the rotor, shifting the supplied power p_{del} towards the right of the maximum power p_{MPP} [30–32]. When the additional power Δp_d is supplied, the rotor speed has to be reduced to ω_{MPP}, releasing kinetic energy [33]. However, despite the fact that this technique can improve the long term frequency regulation, it is not an economically viable solution for wind power plants' operators due to loss of profits [34].

Due to the power inverter, VSWTs cannot naturally provide the kinetic energy stored in their rotor and generator. To overcome this, one or more additional control loops must be included in the power inverter. Three different possibilities can be found in the specific literature: (i) the droop control, (ii) the hidden inertia emulation control, and (iii) the fast power reserve approach. The droop control provides an additional active power Δp proportional to the frequency deviation Δf, following $\Delta p = -\frac{\Delta f}{R_{WT}}$, where R_{WT} is the droop control setting of the VSWT [35–39]. This definition of Δp gives an adaptive response depending on the frequency excursion severity and thus emulating primary frequency control of conventional generation units [40]. The hidden inertia emulation control usually includes two different loops: one considering the RoCoF and the other considering the frequency excursion ($\Delta p \propto$ RoCoF & Δf) [41–43]. However, there are also proposals to use only one additional loop, being $\Delta p \propto$ RoCoF [44–46]. Even though these methods improved the nadir frequency (minimum value), a little frequency dip was observed in later stages. This was due to a small reduction in the generated power compared to the prefault active power (thus, not operating in the MPP) [47]. The fast power reserve approach defines the overproduction power Δp as a constant value independent of the power system configuration and frequency deviation [48–51] or as a variable value depending on the frequency deviation or minimum rotor speed limits [52–54]. With these three techniques, as the additional power Δp is provided, the rotor and generator rotational speeds decrease (subsequently, modifying their torques). Rotor speed variations cause large amplitude edgewise vibrations for the blades [55], affecting the productivity and reducing the efficiency [56]. Large torque increases can address severe mechanical loads on the turbine, even causing critical situations under high mechanical stress conditions [57]. Moreover, consecutive torque increments is related to random load cycles, with important influences on fatigue loads [58].

In this work, a new fast power reserve controller for frequency regulation is proposed for isolated power systems including conventional generation (thermal units) and wind power plants (VSWTs). The proposed adaptive frequency controller is based on a linear regression from different power system parameters (i.e., RoCoF, active power supplied by each synchronous group, and synchronous inertia) to estimate the additional power provided by the wind power plants by maintaining certain frequency thresholds. This way mechanical stress is reduced without excessively prejudicing power system. Subsequently, VSWTs do not always participate in the frequency control but only when they are required according to both the monitored variables. VSWTs frequency control contributions are thus optimized, improving the grid frequency response and providing similar or lower load shedding actions in line with previous frequency control strategies. This proposed VSWT controller approach is tested in the Gran Canaria Island (Canarian archipelago, Spain), an isolated power system where the wind power capacity has doubled since 2017 (from 90 to 180 MW) [59]. Moreover, from 2005 to 2010, more than 200 trips of generators were registered per year in the Canarian archipelago, hence activating the corresponding load shedding programs [60].

The rest of the paper is organized as follows: Section 2 describes the mathematical model used to simulate the power system under consideration; the new frequency control approach proposed in this work is explained in Section 3; cases of study and simulation results are provided in Section 4; Finally, Section 5 gives the conclusions.

2. Power System Modeling

A mathematical model has been designed to analyze the proposed VSWT adaptive frequency control in an isolated power system. A generic wind-thermal isolated system is considered in this study (refer to Figure 2), including steam, diesel, gas, and combined cycle units, as well as wind power plants. Consequently, the model developed in Matlab/Simulink (2016) includes these thermal units, one equivalent VSWT acting as the wind power plants, the automatic generation control (AGC), the power system, and the power demand (including a load shedding scheme). The electromagnetic transients are supposed to be much faster than the other components of the model, and their influence in the system's dynamics is omitted [61]. In Figure 3, the block diagram of the power system under study can be seen.

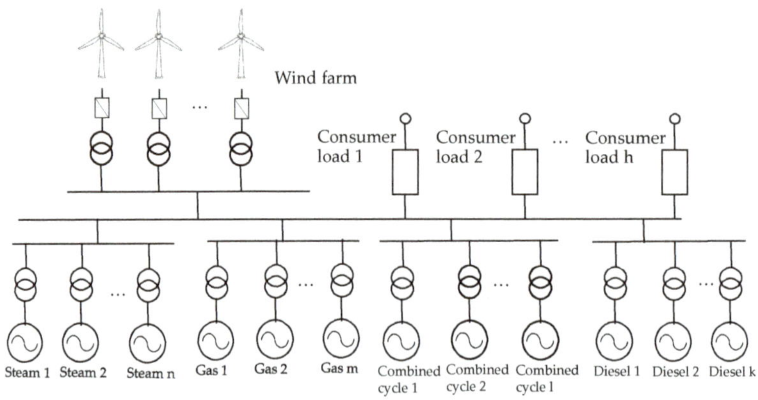

Figure 2. Simplified one-line diagram of the generic isolated power system.

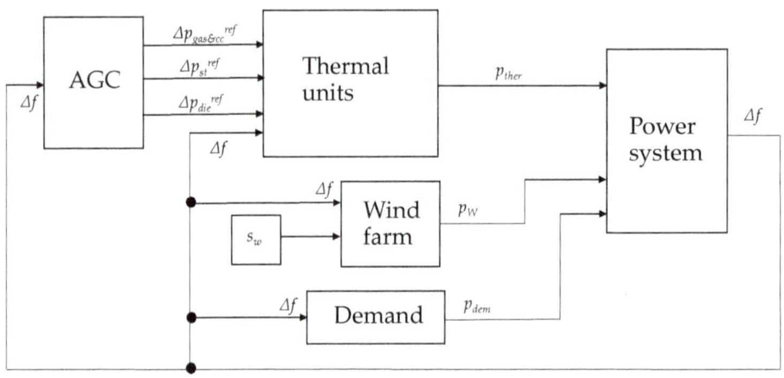

Figure 3. Block diagram of the frequency analysis model.

2.1. Power System

Frequency regulation based on droop speed control can be divided into two consecutive control actions: frequency containment and frequency restoration. On the one hand, frequency containment (or primary frequency regulation), which is based on the governor control, adjusts the active power of the generation units to correct frequency variations. In fact, the frequency nadir is directly related with the generator's droop characteristic. In Spanish electric power system, primary regulation services act up to 30 s after a frequency disturbance [62]. On the other hand, frequency restoration (or secondary frequency regulation) refers to the AGC, adjusting the active power output of generation units to bring system frequency back to its rated value after the governor-based primary response.

An aggregated inertial model is commonly used to analyze frequency deviations in isolated power systems [63]. This modeling approach has been previously applied in El Hierro isolated power system (another isle of the Canarian archipelago) in [64]. Therefore, frequency deviations are the result of the imbalance between the power supplied by the generation units and the power demand:

$$f\,\frac{df}{dt} = \frac{1}{T_{m,ther}(t)}(p_{ther} + p_w - p_{dem} - D_{net}\,\Delta f), \tag{1}$$

where $T_{m,ther}(t)$ corresponds to the total mechanical inertia of thermal units depending on the number of generation operating units at each moment. Note that only thermal units provide inertia to the power system, since VSWTs are decoupled from the grid through power inverters.

When frequency disturbances are higher than certain limits, an under frequency load shedding scheme is activated to recover the grid frequency and fulfill certain frequency range requirements. In this work, a realistic load shedding scheme is included, consisting of the sequential and sudden disconnection of certain amount of load as established frequency thresholds are exceeded. The load shedding program can be found in [59].

2.2. Thermal Power Plants

The different thermal generation technologies (i.e., steam, diesel, gas, and combined cycle) included in the isolated power system are modeled by different transfer functions proposed in [59,65]. These transfer functions supply the power variation of each thermal technology from the frequency deviation and power reference provided by the AGC. Both the model and the parameters can be found in [59]. Since fast response of combined cycle power plants is in charge of the gas turbines, frequency response of combined cycle and gas units is assumed to be equal. The total thermal generation is then the sum of each thermal unit power supplied.

2.3. Variable Speed Wind Turbines

One equivalent VSWT model aggregating all the VSWTs is used [59,66]. The proposed VSWT equivalent model includes the wind power model and both pitch and torque maximum power point tracking control. Further information can be found in [67]. The one-mass rotor mechanical model is used for simulations, which is detailed enough according to [68] for power converters decoupling the generator from the grid. The VSWT diagram is represented in Figure 4. This wind turbine model has been previously used in [69] for short-time period frequency analysis. The equivalent aggregated wind turbine modeling also includes the frequency control response strategy, described in Section 3.

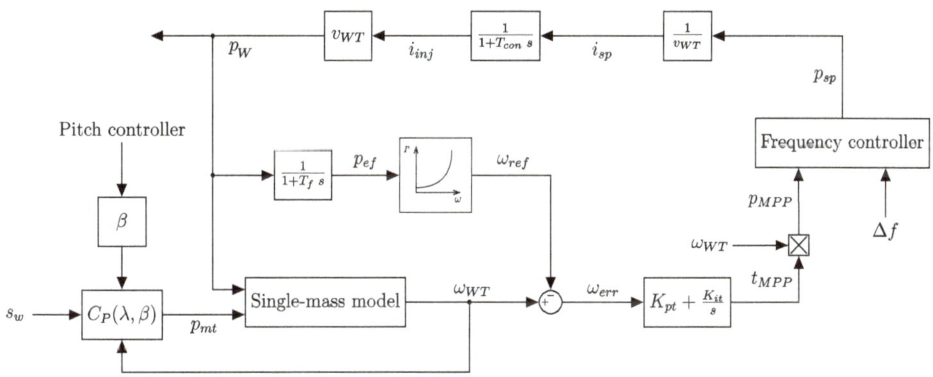

Figure 4. Block diagram of VSWT.

2.4. Automatic Generation Control (AGC)

The secondary control action removes the steady-state frequency error after the primary frequency control. It is modeled similar to [70]. The total secondary regulation effort (ΔRR) is obtained from:

$$\Delta RR = -\Delta f \, K_f, \tag{2}$$

being K_f determined according to the European Network of Transmission System Operators for Electricity (ENTSO-E) recommendations [71]. This regulation effort is distributed among each i synchronized thermal unit as a function of their participation factors ($K_{u,i}$), which are related to the speed droop of each unit [72]. Subsequently, the result of adding all $K_{u,i}$ must be one:

$$\Delta p_i^{ref} = \frac{1}{T_{u,i}} \int \Delta RR \, K_{u,i} \, dt = \frac{-1}{T_{u,i}} \int K_{u,i} \, K_f \, dt, \tag{3}$$

where i represents *gas, cc, die*, and *st* respectively. All thermal units connected to the grid participate in secondary control, whereas the participation factors of those units not connected are considered as zero, $K_{u,i} = 0$.

3. Adaptive Frequency Control Strategy—Methodology

The adaptive control approach proposed in this work tries to minimize the effort of VSWTs when providing frequency response. The proposed approach is based on the fast power reserve technique. This control strategy distinguishes two different periods after a sudden power imbalance: (i) overproduction and (ii) recovery. During overproduction, the stored kinetic energy in the rotating masses of the VSWTs is supplied to the grid as an additional active power Δp during a few seconds, being thus p_w over $p_{MPP}(s_w)$. Subsequently, the rotational speed ω is reduced. Different definitions of this Δp have been proposed. In fact, some authors consider Δp as a fixed constant value [48–51], whereas others propose to estimate Δp as dependent on the torque limit [52] or proportional to the frequency excursion [53]. The recovery period aims to restore ω to the prefault rotational speed value ω_0. To overcome this, p_w should be reduced below $p_{mech}(\omega)$. Previous proposals specified an underproduction power p_{UP} by different ways, being thus the supplied power of each VSWTs: $p_W = p_{mech} - p_{UP}$.

In the proposed adaptive frequency control strategy, the initial value of Δp for the overproduction period is related to the power system conditions. The authors propose a new frequency control approach based on the methodology followed in [73]. This reference estimates the exact and minimum amount of load needed to be shed after an imbalance depending on the RoCoF. In fact, a decision table that links the RoCoF with the strict amount of load shedding is developed based on presimulations. Then, the corresponding load shedding is activated after a contingency, tripping some amount of load demand immediately [74]. In this case, the proposed adaptive controller is based on a decision table that estimates the accurate value of Δp for the VSWT overproduction.

The first step to formulate the decision table consists of defining several simulation scenarios that reflect the variability of the demand, the scheduling units, and/or the wind power penetration. To estimate the overproduction power Δp after the outage of a thermal group, an iterative process is proposed for each scenario (see Figure 5). The condition considered to calculate such Δp is that frequency f should not be below a certain limit f_{lim} for longer than a preset time limit t_{lim}. Both values f_{lim} and t_{lim} are related to the load shedding program of the power system. A counter is thus triggered when f is below f_{lim}, computing the time that frequency is under that f_{lim}. Initially, the i–scenario is simulated assuming that the overproduction power in the first iteration j is equal to 0 ($\Delta p_j = 0$). If f is below f_{lim} for longer than t_{lim}, Δp_j is increased by a fixed value Δp_{inc} with respect to the previous iteration ($\Delta p_j = \Delta p_{j-1} + \Delta p_{inc}$) and the same i–scenario is simulated again with the new value of Δp_j. When the condition is satisfied (i.e., f is below f_{lim} less time that t_{lim}), the minimum Δp for the i–scenario (Δp_i) that VSWTs should provide has been determined ($\Delta p_i = \Delta p_j$). Note that the overproduction power Δp is supplied with a delay of 200 ms, in order to have the measure of the RoCoF and in line with the delay time-interval in between 50 and 500 ms suggested in [75]. Once all the Δp values have been determined from the simulations, a mathematical relationship between such Δp and other variables of the power system need to be found. The obtained expression will be the decision table for the adaptive controller.

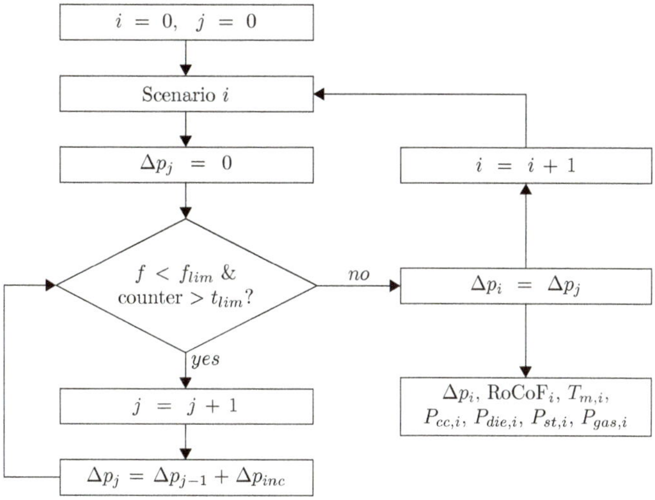

Figure 5. Flow-chart to estimate the overproduction of VSWTs.

When a power imbalance occurs, the VSWTs' controller determines the overproduction power Δp according to the previous decision table. This situation causes a sudden p_w increase and, after that, the supplied power starts to decrease. In this approach, instead of "forcing" the recovery period, the transition to recovery is carried out by the rotor speed PI controller in the converter, which slowly reduces the active power to achieve again $p_{MPP}(s_w)$. In order to avoid a fast power change, that could cause a secondary frequency dip, K_{pt} and K_{it} (proportional and integral constants of the converter, refer to Figure 4) must be conveniently tuned. As a consequence, instead of fixing a p_{UP} or defining an underproduction trajectory, the converter should adapt both the electrical power and the rotational speed to make them return to their pre-event values. Figure 6 presents the evolution of P_W and ω under an imbalanced situation with the proposed approach, pointing out the overproduction and recovery periods.

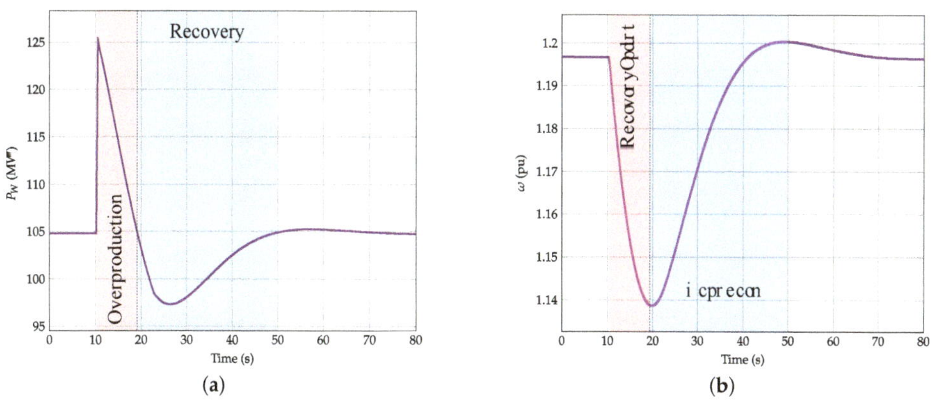

Figure 6. Example of overproduction and recovery periods. (**a**) Electrical power (MW); (**b**) Rotational speed (pu).

4. Case Study—Results

The VSWT adaptive control strategy proposed in this study is applied to the Gran Canaria power system. Gran Canaria Island belongs to Spanish Canarian archipelago, being thus an isolated power system. Its electrical generation has always been based on fossil fuels from two power plants: *Jinámar* and *Barranco de Tirijana* power plants. These power plants included diesel, steam, gas, and combined cycle units. However, in recent decades, the government started to promote wind power plants' installation, doubling its capacity up to 180 MW since 2017. Table 1 lists each thermal unit capacity of the Gran Canaria power system.

Table 1. Gran Canaria thermal units power.

Steam		Gas		Combined Cycle		Diesel	
$P_{st,1}$	72.4 MW	$P_{gas,1}$	32.3 MW	$P_{cc,1}$	206.1 MW	$P_{die,1}$	8.5 MW
$P_{st,2}$	72.4 MW	$P_{gas,2}$	32.3 MW	$P_{cc,2}$	227.0 MW	$P_{die,2}$	8.5 MW
$P_{st,3}$	55.6 MW	$P_{gas,3}$	17.3 MW			$P_{die,3}$	8.5 MW
$P_{st,4}$	55.6 MW	$P_{gas,4}$	32.3 MW			$P_{die,4}$	20.5 MW
		$P_{gas,5}$	32.3 MW			$P_{die,5}$	20.5 MW

4.1. Scenarios under Consideration

Thirty different generation mixes of supply-side programs are under study for different demands and wind power generation. The generation mix scenarios are taken from [59], where a unit commitment model was also included. Figure 7 presents the different supply-side energy schedule of each program. From these generation mix scenarios, two different imbalance conditions are defined: (i) the loss of the largest power plant and (ii) the loss of the second largest power plant. In this way, a total of sixty different scenarios are analyzed. Figure 8 summarizes the supply-side after the disconnection of the largest and second largest units. In addition, it also points out the percentage that represents the loss of each unit over the total system demand.

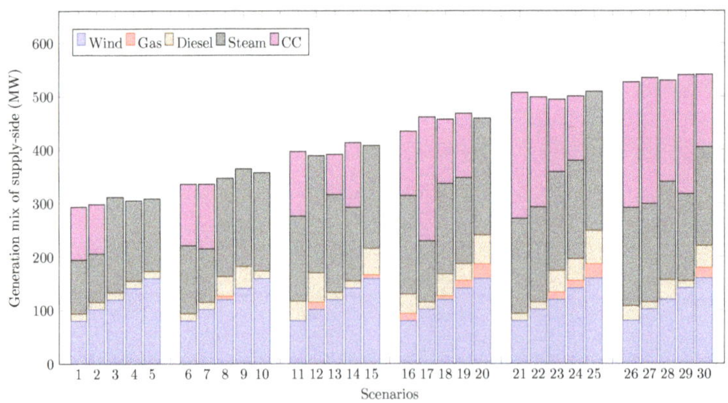

Figure 7. Generation mix scenarios under consideration.

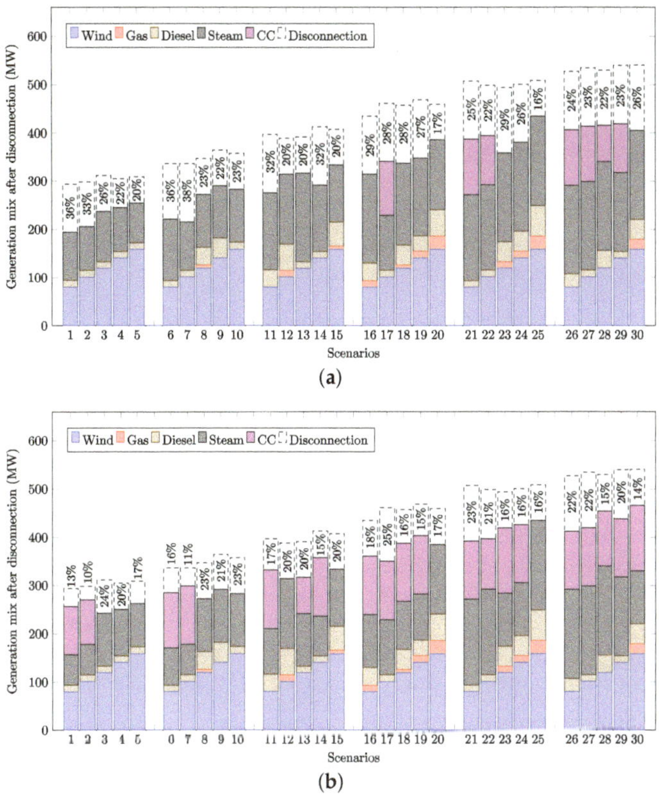

Figure 8. Generation mix after disconnection (**a**) Largest power plant disconnection. (**b**) Second largest power plant disconnection.

4.2. Decision Table Definition: Regression Analysis

From the sixty scenarios presented in Section 4.1, the corresponding Δp values are determined following Section 3. In this work, and according to the load shedding scheme presented in [59], $f_{lim} = 49$ Hz, $t_{lim} = 1$ s and $\Delta p_{inc} = 0.01$ pu for the Gran Canaria island power system. A mathematical relationship between such Δp and other variables of the power system need to be found. As aforementioned, [73] proposed to relate the power to be shed with the RoCoF by a linear and quadratic regression, with $R^2 = 0.951$ and $R^2 = 0.969$, respectively. However, both the linear regression and the quadratic regression of ΔP and RoCoF for this case study accounted for low values of R^2: $R^2 = 0.433$ for linear regression and $R^2 = 0.512$ for quadratic regression. As a consequence, other variables are introduced in the proposed mathematical lineal regression. By also considering the electrical power of each thermal technology assigned in the scheduled program, the system synchronous inertia before the incident, and neglecting all those cases in which the estimated Δp was $\Delta p = 0$, the coefficient of determination increased to $R^2 = 0.801$ following Equation (4):

$$\Delta P = -386.15 + 108.63 \cdot \text{RoCoF} - 1.32 \cdot P_{gas} + 0.26 \cdot P_{die} + 0.65 \cdot P_{st} - 0.42 \cdot P_{cc} + 32.92 \cdot T_m, \quad (4)$$

being ΔP in MW, RoCoF in Hz/s, P_i in MW (i stands for gas, cc, die and st), and T_m in s.

Comparing the results of ΔP obtained from the simulations and those estimated with Equation (4), two additional conditions were included:

1. If the power supplied by the combined cycle power plants before the imbalance was over 180 MW ($P_{cc,0} > 180$ MW), $\Delta P = 0$
2. If the estimated ΔP with the regression is negative ($\Delta P < 0$), $\Delta P = 0$

Moreover, the maximum overproduction power ΔP of the wind power plant was considered as 20% over the installed capacity of the wind power plant, $\Delta P_{max} = 0.2 \cdot P_w$.

4.3. Results

Apart from the adaptive control approach, other frequency strategies have been simulated and compared for the sixty scenarios under consideration:

1. Frequency control only provided by thermal power plants (referred to as *without VSWTs*).
2. Frequency control provided by thermal power plants and VSWTs. The overproduction is defined as proportional to Δf; change from overproduction to recovery period occurs when $\omega/\omega_0 < 0.9$, and $p_{UP} = 0.05$ pu (fast power reserve technique). It is triggered when $\Delta f \geq 100$ mHz and the maximum $\Delta p = 0.15 \cdot p_{mech}(\omega)$ (referred to as *previous approach*).
3. Frequency control provided by thermal power plants and VSWTs using the proposed control approach (referred to as *adaptive approach*).

As was discussed in Section 4.1, sixty different scenarios are considered to evaluate the new frequency control approach for VSWTs. Results are focused on four different aspects for each scenario: (i) the minimum rotational speed ω_{min}; (ii) the maximum torque T_{max}; (iii) the power shed due to the corresponding load shedding programs; (iv) number of cases in which VSWTs do not participate in frequency control with the adaptive proposed approach. Both ω_{min} and T_{max} are expressed in % with regard to the pre-event values, which is equal to their values when they do not participate in frequency control.

Figure 9 depicts the range values of the rotational speed in % compared to the pre-event value. When VSWTs do not participate in frequency control, its ω variation is null, as they are providing $p_{MPP}(s_w)$. According to these results, and depending on the frequency control strategy, important differences can be identified. With regard to the frequency control referred to as *previous approach*, and considering that the change from overproduction to recovery period is always with $\omega/\omega_0 < 0.9$, all scenarios then give the same minimum ω, which represents more than a 7% variation. Based on the *adaptive approach*, ω_{min} variations range in between 0 (for those cases in which the frequency control is not activated) and 6.62%. In fact, the average ω_{min} variation with the proposed strategy is 2.15%. As a consequence, the new frequency control strategy technique does not only optimize those imbalances where VSWTs frequency control participation is required but also reduces the averaged variation of ω to more than half of the previous frequency control approaches.

Figure 10 compares the maximum torque during the VWST frequency response to the corresponding pre-event value. These results are in line with the previous rotational speed comparison, see Figure 9. The torque does not vary when VSWTs do not participate in frequency control, as both ω and p_e remain constant. With the *previous approach* for VSWT frequency control, the maximum torque variation is always the same. Moreover, this variation is over a 30% increase—in comparison to the pre-event value, which can address relevant mechanical loads on the turbine shaft. In contrast, with the proposed *adaptive approach*, the maximum T_{max} variation is 26.5% (only 9 out of 60 scenarios), with an average T_{max} of 9.7%. It can be then deduced that the new frequency control approach does not only avoid the activation of the frequency control of VSWTs in most of imbalance scenarios but also that both ω and T_{max} ranges are reduced significantly compared to previous approaches.

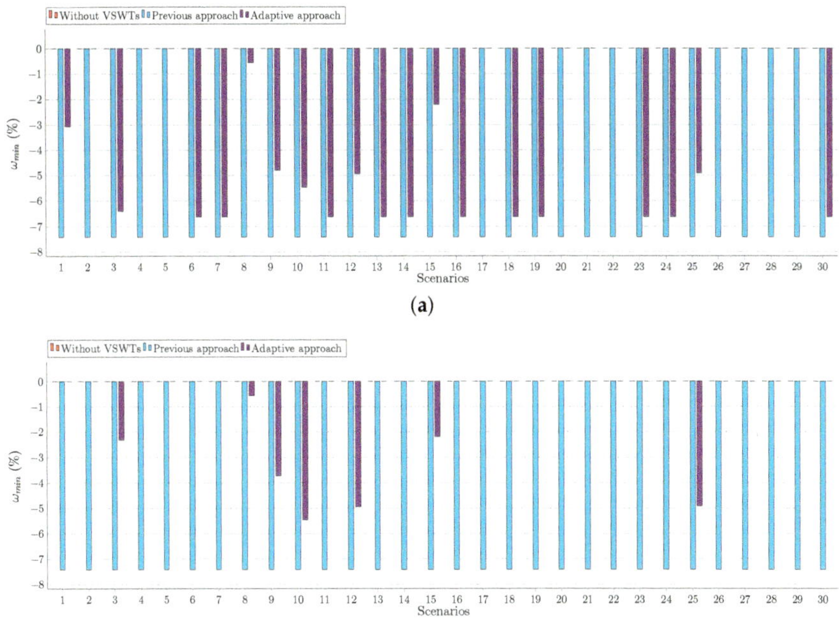

(a)

(b)

Figure 9. Minimum rotational speed (**a**) Largest power plant disconnection; (**b**) Second largest power plant disconnection.

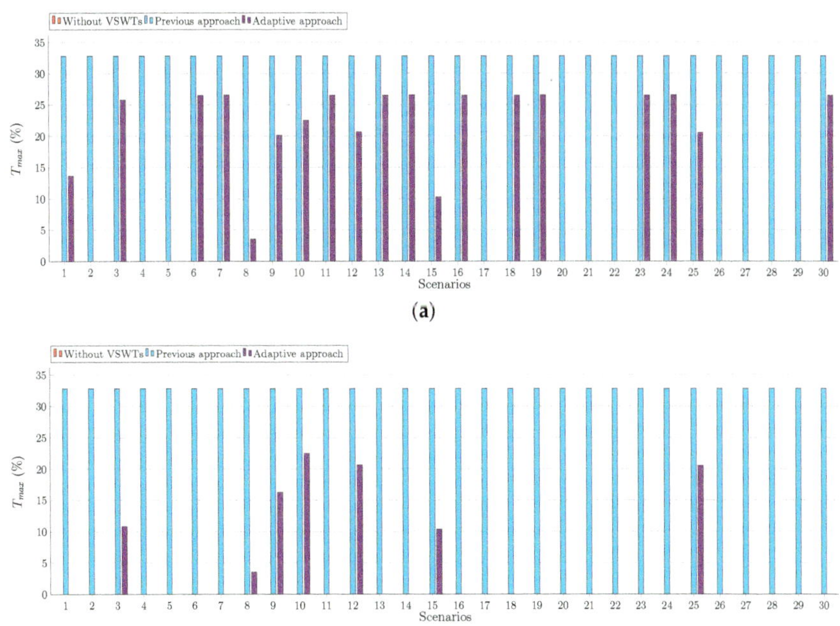

(a)

(b)

Figure 10. Maximum torque variation (**a**) Largest power plant disconnection; (**b**) Second largest power plant disconnection.

Figure 11 shows the power shed by the load shedding program. As can be seen, it is reduced when wind power plants participate in frequency control, compared to the current case in which only thermal power plants are providing such ancillary service. Comparing the two different frequency control approaches analyzed, small differences are found. The *previous approach* gets better results in 4 out of 60 scenarios under study, whereas the *adaptive approach* obtained lower load shedding values in 2 out of 60 scenarios. In the other cases, both frequency controllers obtain the same load shedding.

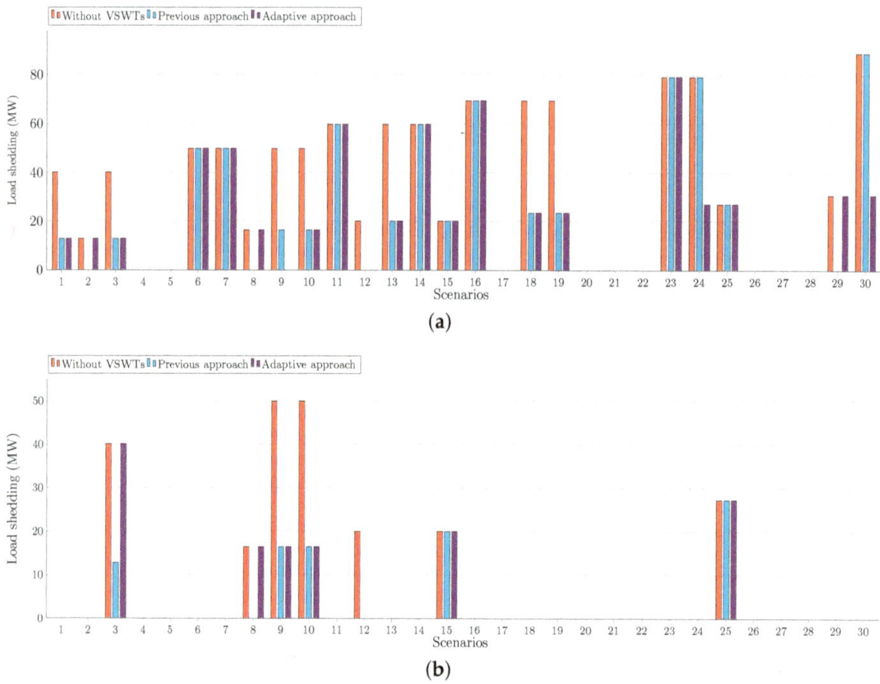

Figure 11. Load shedding (**a**) Largest power plant disconnection; (**b**) Second largest power plant disconnection.

Finally, 57% of imbalances without any VSWTs frequency response participation have been identified. These results reduce significantly the mechanical and electrical VSWT efforts under imbalances, maintaining similar frequency excursions to the previous control strategies. More specifically, the VSWT frequency participation is not required to provide additional power in 11 out of 30 for the largest power plant loss and 23 out of 30 for the second largest power plant loss. These 34 scenarios where the adaptive frequency control is not triggered can be seen in Figures 9 and 10: imbalance scenarios will null variation of ω_{min} and T_{max} for the proposed control strategy. Therefore, p_e and ω values keep as constant during the imbalance and, subsequently, also maintain the torque. In addition, authors have checked that in 94% of the cases in which VSWTs are not participating in frequency control with the *adaptive approach*, the power shed was initially null, pointing out that their participation was not required within acceptable frequency excursion ranges.

As an additional result, Figures 12–14 depict the frequency evolution, load shedding, wind power, and rotational speed for scenarios 9, 16, and 27, respectively, for the loss of the largest power plant. Together with the aforementioned advantages of the new approach, authors also point out that the rotational speed of VSWTs is recovered in a lower time interval than with the previous frequency control approach. Moreover, the transition from overproduction to recovery period is also smoother, avoiding undesirable secondary frequency dips (refer to Figure 12).

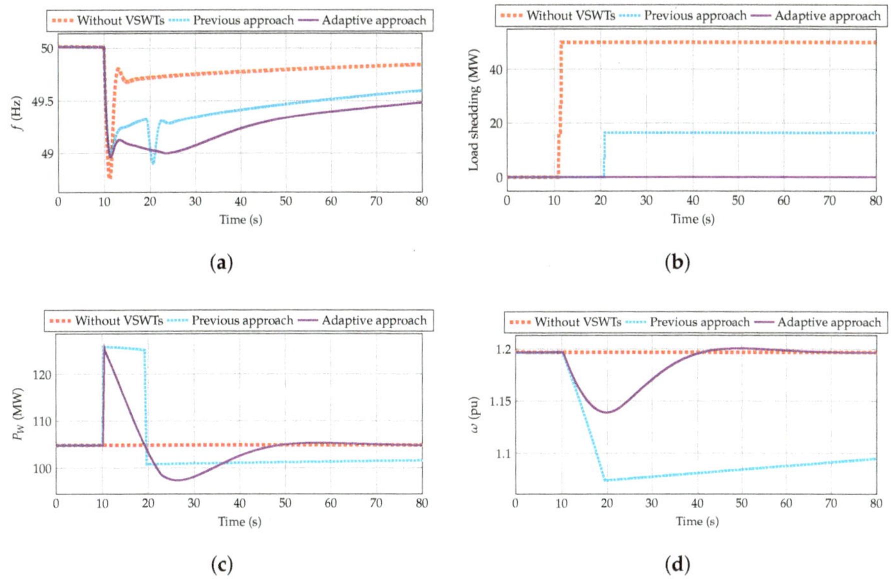

Figure 12. Results for scenario 9 (**a**) Frequency (Hz). (**b**) Load shedding (MW) (**c**) Wind power (MW) (**d**) VSWTs rotational speed (pu).

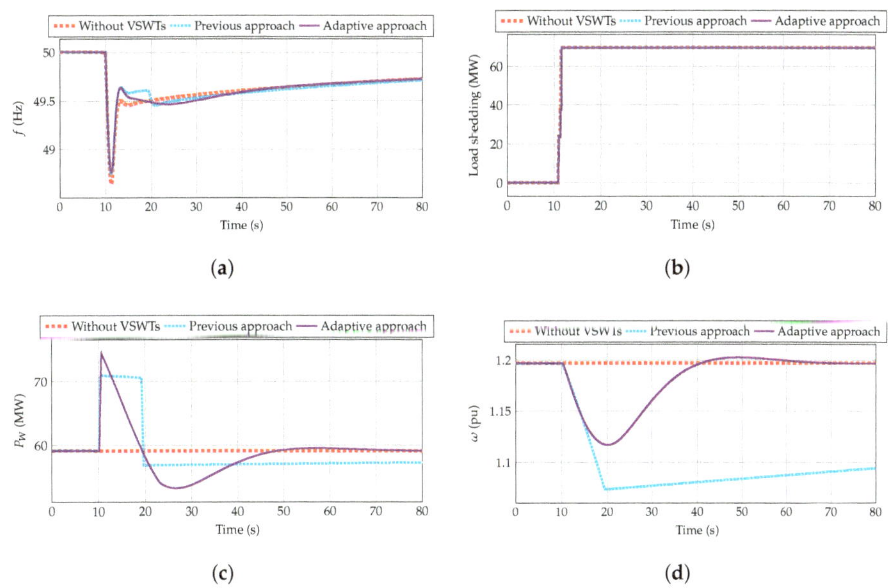

Figure 13. Results for scenario 16 (**a**) Frequency (Hz); (**b**) Load shedding (MW); (**c**) Wind power (MW); (**d**) VSWTs rotational speed (pu).

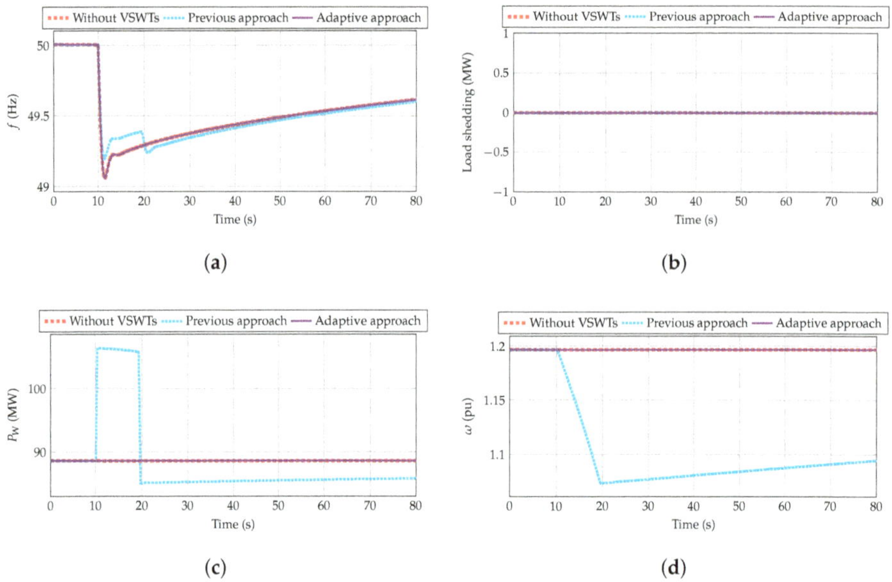

Figure 14. Results for scenario 27 (**a**) Frequency (Hz); (**b**) Load shedding (MW); (**c**) Wind power (MW); (**d**) VSWTs rotational speed (pu).

5. Conclusions

Due to the massive penetration of VSWTs, their contribution to frequency regulation has become a need. As isolated power systems have low synchronous inertia, VSWTs frequency control is even more required. Even though different control strategies for VSWTs have been proposed in the last decade, they usually imply important drawbacks in terms of efficiency, economic profits, and/or equipment wear. As a result, with the aim of reducing the mechanical stress of VSWTs providing frequency response, an adaptive control strategy is proposed in this work. Our approach is based on estimating the minimum overproduction power provided by VSWTs following a linear regression estimation. In this way, such overproduction power depends on some grid parameters (i.e., RoCoF, synchronous inertia, and assigned power of thermal units before the incident). The proposed controller is compared to a conventional fast power reserve strategy. The Gran Canaria isolated power system (Spain) is considered as case study, analyzing sixty representative imbalance scenarios. Results show that similar values for the power shed with the load shedding program are obtained with the two approaches. However, the new adaptive control reduces the VSWTs maximum torque variations and the speed variations (23% and 5% on average, respectively), in comparison to the conventional VSWTs frequency control strategy. This is due to the smoother transition from overproduction to recovery periods, which reduces the electrical and mechanical VSWT efforts. In addition, in 57% of the imbalance scenarios under consideration, VSWTs are not required to participate in frequency response. Thus, both electrical power and rotational speed are kept as constants (subsequently maintaining the torque). As a result, a longer life span is expected for the VSWTs electrical and mechanical components.

Author Contributions: Conceptualization, J.I.S.; Data curation, A.F.-G.; Formal analysis, G.M.-L.; Investigation, A.F.-G.; Methodology, J.I.S.; Resources, A.F.-G. and G.M.-L.; Software, A.F.-G.; Supervision, Á.M.-G. and J.I.S.; Validation, G.M.-L. and J.I.S.; Visualization, Á.M.-G.; Writing—original draft, A.F.-G. and G.M.-L.; Writing—review and editing, A.M.-G. and J.I.S. All authors have read and agreed to the published version of the manuscript.

Funding: This work was partially supported by 'Ministerio de Educación, Cultura y Deporte' of Spain (ref. FPU16/04282) and by 'Ministerio de Economía y Competitividad', under the project "Value of pumped-hydro energy storage in isolated power systems with high wind power penetration" of the National Plan for Scientific and Technical Research and Innovation 2013–2016, grant number ENE2016-77951-R.

Conflicts of Interest: The authors declare no conflict of interest.

Abbreviations

The following abbreviations are used in this manuscript:

β	Pitch angle (grades)
ω	Rotational speed (p.u.)
ΔP	Overproduction power of VSWT (MW)
ΔRR	Total secondary regulation effort
f	Frequency (Hz)
p	Active power (p.u.)
s_w	Wind speed (m/s)
D_{net}	Damping factor of loads
K_u	Participation factor in secondary control
P	Active power (MW)
T_m	Inertia constant (s)
0	Pre-event value (subscript)
cc	Combined cycle (subscript)
dem	Demand (subscript)
die	Diesel (subscript)
gas	Gas (subscript)
max	Maximum value (subscript)
$mech$	Mechanical (subscript)
min	Minimum value (subscript)
st	Steam (subscript)
$ther$	Thermal (subscript)
w	Wind (subscript)
UP	Underproduction (subscript)
AGC	Automatic Generation Control
II-RES	Inverter Interfaced Renewable Energy Sources
MPP	Maximum Power Point
PV	Photovoltaic
RoCoF	Rate of Change of Frequency
TSO	Transmission System Operator
VSWT	Variable Speed Wind Turbine

References

1. Kaushal, A.; Van Hertem, D. An Overview of Ancillary Services and HVDC Systems in European Context. *Energies* **2019**, *12*, 3481. [CrossRef]
2. Martínez-Lucas, G.; Sarasúa, J.I.; Wilhelmi, J.R.; ángel Sánchez, J. Contribution to load-frequency regulation of a hydropower plant with long tail-race tunnel. In Proceedings of the 15th International Conference on Environment and Electrical Engineering (EEEIC), Rome, Italy, 10–13 June 2015; pp. 1886–1891.
3. Qi, Y.; Wang, D.; Wang, X.; Jia, H.; Pu, T.; Chen, N.; Liu, K. Frequency control ancillary service provided by efficient power plants integrated in Queuing-Controlled domestic water heaters. *Energies* **2017**, *10*, 559. [CrossRef]
4. ENTSO-E. Electricity Balancing in Europe. Available online: https://docstore.entsoe.eu/ (accessed on 15 April 2020).
5. Das, K.; Nitsas, A.; Altin, M.; Hansen, A.D.; Sørensen, P.E. Improved load-shedding scheme considering distributed generation. *IEEE Trans. Power Deliv.* **2016**, *32*, 515–524. [CrossRef]

6. Tofis, Y.; Timotheou, S.; Kyriakides, E. Minimal load shedding using the swing equation. *IEEE Trans. Power Syst.* **2016**, *32*, 2466–2467. [CrossRef]
7. Costa, M.H.; Ravetti, M.G.; Saldanha, R.R.; Carrano, E.G. Minimizing undesirable load shedding through robust coordination of directional overcurrent relays. *Int. J. Electr. Power Energy Syst.* **2019**, *113*, 748–757. [CrossRef]
8. Rudez, U.; Mihalic, R. RoCoF-based improvement of conventional under-frequency load Shedding. In Proceedings of the 2019 IEEE Milan PowerTech, Milan, Italy, 23–27 June 2019; pp. 1–5.
9. Fernández-Guillamón, A.; Molina-García, A.; Vigueras-Rodríguez, A.; Gómez-Lázaro, E. Frequency response and inertia analysis in power systems with high wind energy integration. In Proceedings of the 2019 International Conference on Clean Electrical Power (ICCEP), Otranto, Italy, 2–4 July 2019; pp. 388–393.
10. Fernández-Guillamón, A.; Vigueras-Rodríguez, A.; Molina-García, Á. Analysis of power system inertia estimation in high wind power plant integration scenarios. *IET Renew. Power Gener.* **2019**, *13*, 2807–2816. [CrossRef]
11. Molina-García, A.; Fernández-Guillamón, A.; Gómez-Lázaro, E.; Honrubia-Escribano, A.; Bueso, M.C. Vertical Wind Profile Characterization and Identification of Patterns Based on a Shape Clustering Algorithm. *IEEE Access* **2019**, *7*, 30890–30904. [CrossRef]
12. Available online: https://gwec.net/global-wind-report-2019/ (accessed on 29 April 2020).
13. Muñoz-Benavente, I.; Hansen, A.D.; Gómez-Lázaro, E.; García-Sánchez, T.; Fernández-Guillamón, A.; Molina-García, Á. Impact of combined demand-response and wind power plant participation in frequency control for multi-area power systems. *Energies* **2019**, *12*, 1687. [CrossRef]
14. Tielens, P.; Van Hertem, D. The relevance of inertia in power systems. *Renew. Sustain. Energy Rev.* **2016**, *55*, 999–1009. [CrossRef]
15. Daly, P.; Flynn, D.; Cunniffe, N. Inertia considerations within unit commitment and economic dispatch for systems with high non-synchronous penetrations. In Proceedings of the 2015 IEEE Eindhoven PowerTech, Eindhoven, The Netherlands, 29 June–2 July 2015; pp. 1–6.
16. Martínez-Lucas, G.; Sarasúa, J.I.; Sánchez-Fernández, J.Á.; Wilhelmi, J.R. Power-frequency control of hydropower plants with long penstocks in isolated systems with wind generation. *Renew. Energy* **2015**, *83*, 245–255. [CrossRef]
17. Martinez-Lucas, G.; Sarasúa, J.I.; Sánchez-Fernández, J.Á.; Wilhelmi, J.R. Frequency control support of a wind-solar isolated system by a hydropower plant with long tail-race tunnel. *Renew. Energy* **2016**, *90*, 362–376. [CrossRef]
18. Toulabi, M.; Bahrami, S.; Ranjbar, A.M. An input-to-state stability approach to inertial frequency response analysis of doubly-fed induction generator-based wind turbines. *IEEE Trans. Energy Convers.* **2017**, *32*, 1418–1431. [CrossRef]
19. Fernández-Guillamón, A.; Gómez-Lázaro, E.; Muljadi, E.; Molina-García, Á. Power systems with high renewable energy sources: A review of inertia and frequency control strategies over time. *Renew. Sustain. Energy Rev.* **2019**, *115*, 109369. [CrossRef]
20. Alsharafi, A.S.; Besheer, A.H.; Emara, H.M. Primary Frequency Response Enhancement for Future Low Inertia Power Systems Using Hybrid Control Technique. *Energies* **2018**, *11*, 699. [CrossRef]
21. Fernández-Guillamón, A.; Vigueras-Rodríguez, A.; Gómez-Lázaro, E.; Molina-García, Á. Fast power reserve emulation strategy for VSWT supporting frequency control in multi-area power systems. *Energies* **2018**, *11*, 2775. [CrossRef]
22. Chen, Z.; Yin, M.; Zou, Y.; Meng, K.; Dong, Z. Maximum wind energy extraction for variable speed wind turbines with slow dynamic behavior. *IEEE Trans. Power Syst.* **2016**, *32*, 3321–3322. [CrossRef]
23. Ziping, W.; Wenzhong, G.; Tianqi, G.; Weihang, Y.; ZHANG, H.; Shijie, Y.; Xiao, W. State-of-the-art review on frequency response of wind power plants in power systems. *J. Mod. Power Syst. Clean Energy* **2017**, *6*, 1–16.
24. Fernández-Guillamón, A.; Gómez-Lázaro, E.; Muljadi, E.; Molina-Garcia, Á. A Review of Virtual Inertia Techniques for Renewable Energy-Based Generators. In *Power Systems*; IntechOpen: London, UK, 2020.
25. Yingcheng, X.; Nengling, T. Review of contribution to frequency control through variable speed wind turbine. *Renew. Energy* **2011**, *36*, 1671–1677. [CrossRef]
26. Ma, H.; Chowdhury, B. Working towards frequency regulation with wind plants: Combined control approaches. *IET Renew. Power Gener.* **2010**, *4*, 308–316. [CrossRef]

27. Moutis, P.; Papathanassiou, S.A.; Hatziargyriou, N.D. Improved load-frequency control contribution of variable speed variable pitch wind generators. *Renew. Energy* **2012**, *48*, 514–523. [CrossRef]
28. Žertek, A.; Verbič, G.; Pantoš, M. Optimised control approach for frequency-control contribution of variable speed wind turbines. *IET Renew. Power Gener.* **2012**, *6*, 17–23. [CrossRef]
29. Martínez-Lucas, G.; Sarasúa, J.I.; Pérez-Díaz, J.I.; Martínez, S.; Ochoa, D. Analysis of the Implementation of the Primary and/or Inertial Frequency Control in Variable Speed Wind Turbines in an Isolated Power System with High Renewable Penetration. Case Study: El Hierro Power System. *Electronics* **2020**, *9*, 901. [CrossRef]
30. Castro, L.M.; Fuerte-Esquivel, C.R.; Tovar-Hernández, J.H. Solution of power flow with automatic load-frequency control devices including wind farms. *IEEE Trans. Power Syst.* **2012**, *27*, 2186–2195. [CrossRef]
31. Vidyanandan, K.; Senroy, N. Primary frequency regulation by deloaded wind turbines using variable droop. *IEEE Trans. Power Syst.* **2013**, *28*, 837–846. [CrossRef]
32. Zhang, X.; Zha, X.; Yue, S.; Chen, Y. A Frequency Regulation Strategy for Wind Power Based on Limited Over-Speed De-Loading Curve Partitioning. *IEEE Access* **2018**, *6*, 22938–22951. [CrossRef]
33. Wang, S.; Tomsovic, K. A Novel Active Power Control Framework for Wind Turbine Generators to Improve Frequency Response. *IEEE Trans. Power Syst.* **2018**, *33*, 6579–6589. [CrossRef]
34. Fang, X.; Krishnan, V.; Hodge, B.M. Strategic offering for wind power producers considering energy and flexible ramping products. *Energies* **2018**, *11*, 1239. [CrossRef]
35. Ye, H.; Pei, W.; Qi, Z. Analytical modeling of inertial and droop responses from a wind farm for short-term frequency regulation in power systems. *IEEE Trans. Power Syst.* **2016**, *31*, 3414–3423. [CrossRef]
36. Arani, M.F.; Mohamed, Y.A.I. Dynamic Droop Control for Wind Turbines Participating in Primary Frequency Regulation in Microgrids. *IEEE Trans. Smart Grid* **2017**, *9*, 5742–5751. [CrossRef]
37. Lertapanon, P.; Wangdee, W. Analysis and modeling of wind turbine generators considering frequency controls. In Proceedings of the Electrical Engineering Congress (iEECON), Pattaya, Thailand, 8–10 March 2017; pp. 1–4.
38. Huang, L.; Xin, H.; Zhang, L.; Wang, Z.; Wu, K.; Wang, H. Synchronization and Frequency Regulation of DFIG-Based Wind Turbine Generators With Synchronized Control. *IEEE Trans. Energy Convers.* **2017**, *32*, 1251–1262. [CrossRef]
39. Deepak, M.; Abraham, R.J.; Gonzalez-Longatt, F.M.; Greenwood, D.M.; Rajamani, H.S. A novel approach to frequency support in a wind integrated power system. *Renew. Energy* **2017**, *108*, 194–206. [CrossRef]
40. Margaris, I.D.; Papathanassiou, S.A.; Hatziargyriou, N.D.; Hansen, A.D.; Sorensen, P. Frequency control in autonomous power systems with high wind power penetration. *IEEE Trans. Sustain. Energy* **2012**, *3*, 189–199. [CrossRef]
41. Morren, J.; de Haan, S.W.H.; Kling, W.L.; Ferreira, J.A. Wind turbines emulating inertia and supporting primary frequency control. *IEEE Trans. Power Syst.* **2006**, *21*, 433–434. [CrossRef]
42. Díaz-González, F.; Hau, M.; Sumper, A.; Gomis-Bellmunt, O. Participation of wind power plants in system frequency control: Review of grid code requirements and control methods. *Renew. Sustain. Energy Rev.* **2014**, *34*, 551–564. [CrossRef]
43. Dreidy, M.; Mokhlis, H.; Mekhilef, S. Inertia response and frequency control techniques for renewable energy sources: A review. *Renew. Sustain. Energy Rev.* **2017**, *69*, 144–155. [CrossRef]
44. Ekanayake, J.; Jenkins, N. Comparison of the response of doubly fed and fixed-speed induction generator wind turbines to changes in network frequency. *IEEE Trans. Energy Convers.* **2004**, *19*, 800–802. [CrossRef]
45. Gonzalez-Longatt, F.; Chikuni, E.; Stemmet, W.; Folly, K. Effects of the synthetic inertia from wind power on the total system inertia after a frequency disturbance. In Proceedings of the Power Engineering Society Conference and Exposition in Africa, Johannesburg, South Africa, 9–13 July 2012; pp. 9–13.
46. Bonfiglio, A.; Invernizzi, M.; Labella, A.; Procopio, R. Design and Implementation of a Variable Synthetic Inertia Controller for Wind Turbine Generators. *IEEE Trans. Power Syst.* **2018**, *34*, 754–764. [CrossRef]
47. Singarao, V.Y.; Rao, V.; Harral, M.A. Review on engineering and regulatory aspects associated with frequency control capabilities of wind power plants. In Proceedings of the 2013 IEEE Energytech, Cleveland, OH, USA, 21–23 May 2013; pp. 1–6.
48. Tarnowski, G.C.; Kjar, P.C.; Sorensen, P.E.; Ostergaard, J. Variable speed wind turbines capability for temporary over-production. In Proceedings of the Power & Energy Society General Meeting, Calgary, AB, Canada, 26–30 July 2009; pp. 1–7.

49. Keung, P.K.; Li, P.; Banakar, H.; Ooi, B.T. Kinetic energy of wind-turbine generators for system frequency support. *IEEE Trans. Power Syst.* **2009**, *24*, 279–287. [CrossRef]
50. El Itani, S.; Annakkage, U.D.; Joos, G. Short-term frequency support utilizing inertial response of DFIG wind turbines. In Proceedings of the Power and Energy Society General Meeting, Detroit, MI, USA, 24–28 July 2011; pp. 1–8.
51. Hafiz, F.; Abdennour, A. Optimal use of kinetic energy for the inertial support from variable speed wind turbines. *Renew. Energy* **2015**, *80*, 629–643. [CrossRef]
52. Kang, M.; Kim, K.; Muljadi, E.; Park, J.W.; Kang, Y.C. Frequency control support of a doubly-fed induction generator based on the torque limit. *IEEE Trans. Power Syst.* **2016**, *31*, 4575–4583. [CrossRef]
53. Fernández-Guillamón, A.; Villena-Lapaz, J.; Vigueras-Rodríguez, A.; García-Sánchez, T.; Molina-García, Á. An Adaptive Frequency Strategy for Variable Speed Wind Turbines: Application to High Wind Integration Into Power Systems. *Energies* **2018**, *11*, 1436. [CrossRef]
54. Liu, K.; Qu, Y.; Kim, H.M.; Song, H. Avoiding Frequency Second Dip in Power Unreserved Control During Wind Power Rotational Speed Recovery. *IEEE Trans. Power Syst.* **2018**, *33*, 3097–3106. [CrossRef]
55. Staino, A.; Basu, B. Dynamics and control of vibrations in wind turbines with variable rotor speed. *Eng. Struct.* **2013**, *56*, 58–67. [CrossRef]
56. Rahman, M.; Ong, Z.C.; Chong, W.T.; Julai, S.; Khoo, S.Y. Performance enhancement of wind turbine systems with vibration control: A review. *Renew. Sustain. Energy Rev.* **2015**, *51*, 43–54. [CrossRef]
57. Hansen, A.D.; Altin, M.; Margaris, I.D.; Iov, F.; Tarnowski, G.C. Analysis of the short-term overproduction capability of variable speed wind turbines. *Renew. Energy* **2014**, *68*, 326–336. [CrossRef]
58. Lind, P.G.; Herráez, I.; Wächter, M.; Peinke, J. Fatigue load estimation through a simple stochastic model. *Energies* **2014**, *7*, 8279–8293. [CrossRef]
59. Fernández-Guillamón, A.; Sarasúa, J.I.; Chazarra, M.; Vigueras-Rodríguez, A.; Fernández-Muñoz, D.; Molina-García, Á. Frequency control analysis based on unit commitment schemes with high wind power integration: A Spanish isolated power system case study. *Int. J. Electr. Power Energy Syst.* **2020**, *121*, 106044. [CrossRef]
60. Padrón, S.; Hernández, M.; Falcón, A. Reducing under-frequency load shedding in isolated power systems using neural networks. Gran Canaria: A case study. *IEEE Trans. Power Syst.* **2015**, *31*, 63–71. [CrossRef]
61. Martínez-Lucas, G.; Sarasúa, J.I.; Sánchez-Fernández, J.Á. Frequency regulation of a hybrid wind–hydro power plant in an isolated power system. *Energies* **2018**, *11*, 239. [CrossRef]
62. Aziz, A.; Oo, A.T.; Stojcevski, A. Frequency regulation capabilities in wind power plant. *Sustain. Energy Technol. Assess.* **2018**, *26*, 47–76. [CrossRef]
63. Mansoor, S.; Jones, D.; Bradley, D.A.; Aris, F.; Jones, G. Reproducing oscillatory behaviour of a hydroelectric power station by computer simulation. *Control Eng. Pract.* **2000**, *8*, 1261–1272. [CrossRef]
64. Sarasúa, J.I.; Martínez-Lucas, G.; Lafoz, M. Analysis of alternative frequency control schemes for increasing renewable energy penetration in El Hierro Island power system. *Int. J. Electr. Power Energy Syst.* **2019**, *113*, 807–823. [CrossRef]
65. Kundur, P.; Balu, N.J.; Lauby, M.G. *Power System Stability and Control*; McGraw-Hill: New York, NY, USA, 1994; Volume 7.
66. Pöller, M.; Achilles, S. Aggregated wind park models for analyzing power system dynamics. In Proceedings of the 4th International Workshop on Large-Scale Integration of Wind Power and Transmission Networks for Offshore Wind Farms, Billund, Denmark, 20–21 October 2003.
67. Clark, K.; Miller, N.W.; Sanchez-Gasca, J.J. Modeling of GE wind turbine-generators for grid studies. *GE Energy* **2010**, *4*, 0885–8950.
68. Zhao, S.; Nair, N.K. Assessment of wind farm models from a transmission system operator perspective using field measurements. *IET Renew. Power Gener.* **2011**, *5*, 455–464. [CrossRef]
69. Martínez-Lucas, G.; Sarasúa, J.I.; Sánchez-Fernández, J.Á. Eigen analysis of wind–hydro joint frequency regulation in an isolated power system. *Int. J. Electr. Power Energy Syst.* **2018**, *103*, 511–524. [CrossRef]
70. Perez-Diaz, J.I.; Sarasua, J.I.; Wilhelmi, J.R. Contribution of a hydraulic short-circuit pumped-storage power plant to the load–frequency regulation of an isolated power system. *Int. J. Electr. Power Energy Syst.* **2014**, *62*, 199–211. [CrossRef]

71. UCTE (Union for the Co-ordination of Transmission of Electricity). Operation Handbook. 2004. Available online: https://www.entsoe.eu/publications/system-operations-reports/operation-handbook (accessed on 4 May 2020).
72. Wood, A.J.; Wollenberg, B.F.; Sheblé, G.B. *Power Generation, Operation, and Control*; John Wiley & Sons: Hoboken, NJ, USA, 2013.
73. Sigrist, L. A UFLS scheme for small isolated power systems using rate-of-change of frequency. *IEEE Trans. Power Syst.* **2014**, *30*, 2192–2193. [CrossRef]
74. Dai, Y.; Xu, Y.; Dong, Z.; Wong, K.; Zhuang, L. Real-time prediction of event-driven load shedding for frequency stability enhancement of power systems. *IET Gener. Transm. Distrib.* **2012**, *6*, 914–921. [CrossRef]
75. Arani, M.F.M.; Mohamed, Y.A.R.I. Analysis and mitigation of undesirable impacts of implementing frequency support controllers in wind power generation. *IEEE Trans. Energy Convers.* **2015**, *31*, 174–186. [CrossRef]

energies

MDPI

Article

Flywheel Energy Storage and Dump Load to Control the Active Power Excess in a Wind Diesel Power System

Rafael Sebastián [1,*] and Rafael Peña-Alzola [2]

1 Department of Electrical, Electronic and Control Engineering (DIEEC), UNED, 28040 Madrid, Spain
2 Technology and Innovation Centre, University of Strathclyde, Glasgow G1 1RD, Scotland, UK; rafael.pena-alzola@strath.ac.uk
* Correspondence: rsebastian@ieec.uned.es; Tel.: +34-913987624

Received: 9 March 2020; Accepted: 13 April 2020; Published: 18 April 2020

Abstract: Wind Diesel Power Systems (WDPS) are isolated microgrids which combine Wind Turbine Generators (WTGs) with Diesel Generators (DGs). The WDPS modelled in this article is composed of a DG, a WTG, consumer load, Dump Load (DL) and a Flywheel Energy Storage System (FESS). In the Wind-Diesel (WD) mode both the DG and WTG supply power to the consumers. The WDPS is simulated in the WD mode in the case that the WTG produced power exceeds the load consumption. This WTG excess power case is simulated in the subcases of DL and FESS turned off, only-DL and only-FESS. Simulations for the DL and FESS-off case show that the WTG excess power leads to a continuous system frequency increase, so that the tripping of the WTG Circuit Breaker (CB) is required to guarantee the WDPS power supply continuity. Simulations for the only-DL/only-FESS cases show that commanding the DL/FESS to consume controlled power, so that the required DG power to balance the system active power is positive, enables the DE speed governor to regulate the system frequency. Furthermore, the frequency and voltage variations in the DL/FESS cases are moderate and there is no need to trip the WTG-CB, so that the WDPS reliability and power quality are greatly improved. Additionally, the only-FESS case obtains better WDPS relative stability than the only-DL case.

Keywords: diesel generator; wind turbine generator; isolated microgrid; flywheel energy storage; dump load; power systems simulation; power systems control

1. Introduction

Wind diesel power systems (WDPS) are isolated microgrids which combine wind turbine generators (WTGs) with diesel generators (DGs) to supply electrical power to remote consumers. All WDPSs have two modes of operation [1]: Diesel-Only (DO) mode, where the DGs supply all the power (active and reactive) to the isolated consumers, and Wind-Diesel (WD) mode, where both the WTGs and the DGs supply active power. In both DO and WD modes, the system frequency regulation is performed by speed governors included in each diesel engine (DE) and system voltage regulation is performed by automatic voltage regulators (AVR) included in each synchronous machine (SM). High penetration WDPSs can also work in Wind-Only (WO) mode, where the WTGs are the only active power suppliers and the DGs do not run.

WDPSs are low inertia isolated power systems, where the balance between power generation and consumption is difficult to achieve due to the uncontrolled WTGs power production and consumer loading. As a result, the WDPS frequency and voltage can have significant deviations. The WDPS stability and power quality have been dealt with in literature mostly by the dynamic simulation of different WDPS architectures. Ref. [2] shows how the variations in load and WTG power affects the power quality of a no-storage WDPS. In the no-storage WDPS of ref. [3], a static reactive compensator and a synchronous machine voltage regulator are coordinated to control the system voltage. In the WDPS of [4], distributed resistive loads are controlled to support the speed governor of the diesel engine in the regulation of frequency. When a short-term Energy Storage System (ESS) is added to a WDPS, several benefits such as voltage and frequency support and increasing stability [5] are obtained. Previous benefits are greater in WDPS than in large power systems that have much bigger inertia. The WDPS frequency in [6] is stabilized by an ultra-capacitor based ESS. In [7] a battery-based ESS (BESS) in a WDPS supplies the active power needed to prevent temporarily a DG overloaded situation, so that load shedding is avoided and the WDPS reliability is increased. In [8] the simulations in WD mode of a WDPS with a flywheel based ESS (FESS) show that the WDPS power quality is improved by the addition of the FESS.

WDPS are isolated microgrids and the following microgrid studies are related to WDPS: [9] shows a BESS providing frequency support to a microgrid with high penetration of renewable energy sources, and the BESS in [9] is also used as an uninterrupted power supply for critical loads, a working mode that can be used in the ESS employed in a WDPS; in [10], a BESS supports voltage regulation by counteracting the voltage variations resulting from power fluctuations of renewable power sources; ref. [11] shows how a BESS included in a DGs-based isolated ship power plant smooths the active power variation, so that this study can be applied to the DO mode of a WDPS; ref. [12] shows, with simulation results, the coordination between a supercapacitor and a battery ESSs to balance the active power in an isolated microgrid with only a WTG as a generator. The simulations in [12] can be applied to the WO mode of a WDPS. The "El Hierro" island Diesel-Hydro-Wind power system includes hydropower pumped-storage and has been simulated with the DGs shut-off (Wind-Hydro mode) in the following publications: ref [13] shows the system frequency regulation by using the variable and fixed speed pumps integrated into the hydropower pumped-storage; ref [14] shows, among other frequency control schemes, how the frequency regulation is improved by adding a FESS to the system.

The first part of this article presents the modelling of the WDPS shown in Figure 1, which can work in DO and WD modes. In addition to a DG, WTG and consumer load, the WDPS includes a dump load (DL) and a flywheel ESS (FESS). The second part and main goal of this article aims to present solutions for the WD mode of the WDPS of Figure 1 in the case where the WTG produced power P_T exceeds the load consumption P_L, a situation that makes the WDPS unstable. The WTG power excess situation ($P_T > P_L$) is simulated in WD mode in three cases, namely DL and FESS-off (DL and FESS are turned off), only-DL (DL actuates but FESS is turned off) and only-FESS (FESS actuates but DL is turned off). In the DL and FESS-off simulation case, it is shown that to prevent the isolated power system collapse that the WTG power excess provokes, the solution is to trip the WTG circuit breaker (I_T in Figure 1). In both the only-DL and only-FESS simulation cases, it is shown how the DL/FESS are commanded to consume controlled power, avoiding the WTG circuit breaker trip necessary in the DL and FESS-off case so that the WDPS absolute stability is increased. A previous paper [15] deals with simulations in WO mode of a system that comprises a WTG, a grid forming SM, load and a FESS. Ref [15] does not include a diesel engine, therefore this article's isolated power system architecture is different. In a previous paper [8], the simulated WDPS includes a FESS, but no DL is considered. Furthermore, the WTG power levels in [8] are below the load consumption ($P_T < P_L$) in all the simulations, so [8] does not deal with the WTG power excess situation. In addition, [8] focuses on the control of the FESS power converter. Previous papers [5] and [7] use a battery ESS instead of a flywheel ESS, so the used power converter and ESS variables shown in the simulations are different. Additionally, [5] and [7] do not use the DL. The WD mode simulations in [7] do not consider the WTG power excess situation dealt in this paper. Ref. [5] deals with a high penetration WDPS and its main simulations are in WO mode.

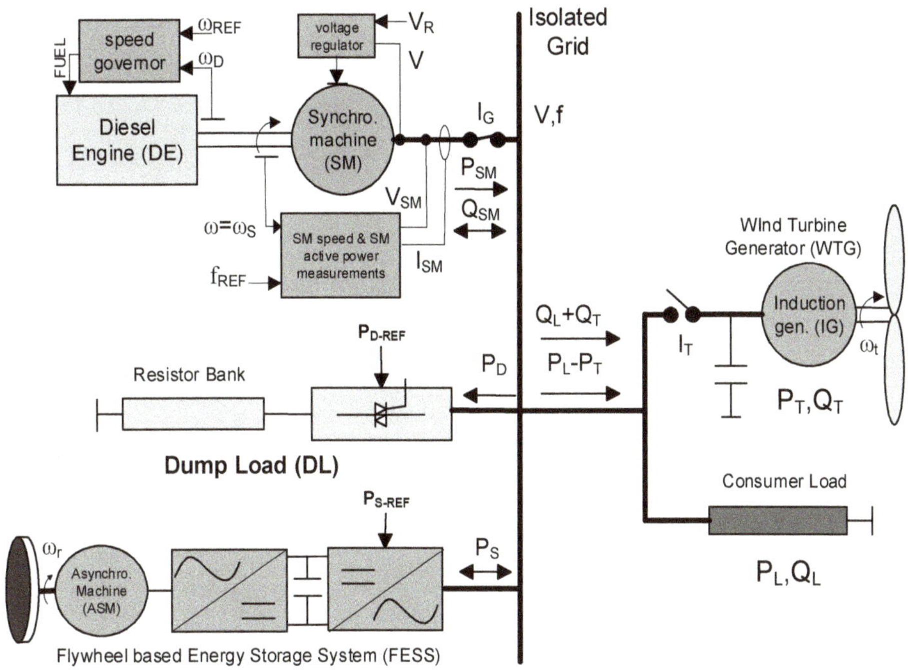

Figure 1. Layout of the Wind Diesel Power System (WDPS) with Dump Load (DL) and Flywheel Energy Storage.

After this introduction, this article contains the following sections: Section 2 presents the modelling of the WDPS components: DG, WTG, DL and FESS. The WTG power excess situation is analysed and simulated in Section 3 for the DL and FESS-off case and in Section 4 for the only-DL and only-FESS cases. Section 4 also compares its simulation results with the ones in Section 3 and makes a comparison between the only-DL and only-FESS cases. The last section contains conclusions, summarizing the benefits in terms of greater stability and reliability of using the DL/FESS.

2. Isolated WDPS Modelling

Figure 1 shows the WDPS presented in this paper. All the parameters of the presented WDPS are shown in Appendix A. The DG supplies controlled active and reactive power and the WTG supplies uncontrolled active power. The consumer load consumes uncontrolled active power and the DL consumes controlled active power. The FESS consumes/generates controlled active power. The state opened/closed of the WTG circuit breaker I_T defines the DO/WD operation mode of the WDPS.

The MATLAB–Simulink framework was used to model and simulate the WDPS, and Figure 2 shows the WPDS Simulink schematic. The next subsections describe the modelling of the WDPS components.

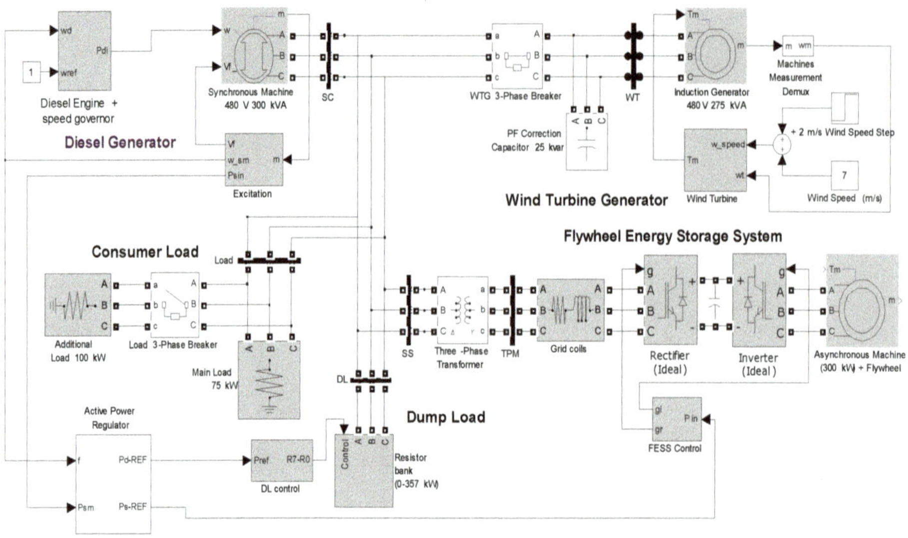

Figure 2. Simulink-SimPowerSystem of the WPDS with Dump Load and Flywheel Energy Storage.

2.1. The Diesel Generator Model.

The DG consists of a Diesel Engine (DE) and a Synchronous machine (SM), and its rated power is 300 kVA. The DE converts the fuel energy into the shaft mechanical power P_D and the SM converts the DE mechanical power into electrical power. The SM provides the isolated grid sinusoidal voltage waveform. By following the command of its automatic voltage regulator, the SM provides reactive power to the WDPS, which is necessary to keep the system voltage module V within the allowable limits. The relationship between the voltage waveform frequency f (Hz) and the shaft speed of the DE/SM ω (rad/s) is:

$$\omega = 2\pi f/p \tag{1}$$

where p is the SM number of pole pairs.

To control the system frequency, a speed governor controls the speed of the DE. The speed governor comprises a speed regulator and an actuator. In this article, the speed regulator applies a Proportional Integral derivative (PID) type algorithm to the DE speed error ($\omega_{ref} - \omega_d$ in Figure 2, where ω_{ref} is the DE reference speed and ω_d the actual speed ω) so the DE speed control is isochronous, that is, in steady state the DE speed is rated one (therefore rated system frequency) provided that the electrical load is within 0–300 kW range. The actuator converts the output of the speed regulator into a proper signal to control a fuel valve. In this way, the incoming fuel rate to the DE is adjusted to control the DE produced mechanical power to the needed value to achieve the rated speed in the DE.

In this article, the SM electrical part is modelled by a sixth-order model and the SM automatic voltage regulator model is the IEEE type 1. The SymPowerSystem blockset [16] provides both models.

The models for the DE and its speed regulator described in [17] are followed. The DE model consists of a gain which relates the DE mechanical power with the rate of fuel consumed and a transport lag to model the firing delay between pistons. A second order system models the actuator. The DG inertia constant H_{DG} is 1.75 s.

2.2. The Wind Turbine Generator Model

The 275 kW WTG comprises a fixed pitch Wind Turbine (WT) driving through a gearbox a Squirrel Cage Induction Generator (SCIG). The WT converts the wind power into shaft mechanical power and the SCIG converts the WT mechanical power into electrical power. The fixed pitch WT model follows the one in ref. [18] and is included in the Wind Turbine block of Figure 2. The model consists of the wind turbine power curves that relate the mechanical shaft power produced by the WT $P_{T\text{-}MEC}$ with the wind speed (v_speed) and the WT shaft speed ω_t. The output of the Wind Turbine block is the torque applied to the SCIG ($T_m = P_{T\text{-}MEC}/\omega_t$)

For generator operation, the SCIG speed range is very limited, within a narrow range between 1 and 1.02 of the synchronous speed [19], and for this reason, this WT-SCIG type is called constant speed WTG. Therefore this WT-SCIG type does not allow to adjust its rotational speed to maximize the capture of wind energy [20]. Other types of WTGs used in WDPS [21–24] allow variable speed operation to maximize the capture of wind energy by performing a maximum power point tracking technique. In [12,21–23], the WTG equips a synchronous generator and an AC-DC-AC electronic power converter, which connects the synchronous generator to the grid. In [24], the WTG equips a double fed IG with its stator connected to the grid and its rotor connected to an AC-DC-AC converter through a slip ring. Additionally, as the used WT is a fixed pitch blade type, $P_{T\text{-}MEC}$ is mainly a function of the cube of the wind speed [25]. The wind speed is quasi-random, so that the WT-SCIG behaves as an uncontrolled active power supplier. In spite of the previously commented disadvantages, the WT-SCIG fixed pitch constant speed type used in this article has remarkable features for the remote locations of WDPS, such as robust construction, simple maintenance and low cost. The used WT-SCIG type is more robust than [12,21–24] as it does not equip an electronic power converter and has less maintenance than [24], as the SCIG does not have slip rings.

The WT-SCIG has two additional positive features. First, as the SCIG stator is directly connected to the isolated grid, the WTG inertia also participates in the moderation of the WDPS frequency changes. This property is not available in [12,21–23] as the double power converter uncouples the WTG- electrical generator from the grid. Second, as in the WTG-SCIG working speed range the SCIG torque is proportional to its slip ($\omega_t - \omega$ in per unit values) [25], the WTG-SCIG improves the system frequency by providing a damped response [26].

The 25 kVA capacitor bank of Figure 2 provides reactive power to improve the WTG-SCIG power factor. The SymPowerSystem blockset [16] provides the SCIG model. The SCIG electrical model is a fourth-order one. Typical WTG inertia constants H_W range between 2–6 s [27] for low–high power WTGs. The used WTG is a low power one, so that H_W = 2 s is set for this parameter.

2.3. *The Dump Load Model*

The DL in a WDPS is actuated when a WTG power excess exists. In the WD mode, the WDPS control commands the DL to dump the power needed to ensure a DG minimum positive load, avoiding a DG reverse power. In the WO mode, the WDPS control commands the DL to consume the WTG power excess (P_T-P_L) to regulate system frequency [23,28]. The Figure 2 DL consists of eight three-phase resistors connected in series with solid-state switches, which connect/disconnect the resistors at the zero crossing in order to prevent harmonic injection. The values of the resistors follow a binary progression, with values R, $R/2$, $R/2^2$ … $R/2^7$. If P_0 is the rated power of the resistor of value R ($P_0 = V_n{}^2/R$, where V_n is the rated system voltage), the eight resistor's rated powers are P_0, $2{\cdot}P_0$, $2^2{\cdot}P_0$, … , $2^7{\cdot}P_0$. Being I_J the state closed(1)/opened(0) of the three phase switch associated with the resistor of power $2^J{\cdot}P_0$, the power consumed by the DL P_D if the system voltage is the rated value, can be expressed as:

$$P_D = (I_0 + I_1{\cdot}2^1 + \ldots + I_7{\cdot}2^7){\cdot}P_0 \tag{2}$$

According to Equation (2), P_D can be varied discretely in steps of P_0 from 0 to 255$\cdot$ P_0. P_0 was chosen to be 1.4 kW, so the DL rated power $P_{D\text{-}NOM}$ is 357 kW (1.4·255). The DL control block of Figure 2 determines which resistors have to be connected to consume the assigned $P_{D\text{-}REF}$. The details of the DL implementation in Simulink can be found in [18].

2.4. The Flywheel Energy StorageSystem Model

A low speed (LS) FESS type [15] was included as a short term ESS in the WDPS. The LS-FESS consists of a steel flywheel driven by an electrical asynchronous machine (ASM). Features such as high power/torque, robust construction, low cost and wide availability are determinant to select an ASM for WDPS. The simple manufacturing and low cost are determinant to select a steel flywheel [29]. The rotating flywheel stores kinetic energy. The ASM converts the kinetic energy into electrical energy and the electrical energy into kinetic energy when it works as a generator/motor respectively. The flywheel has the moment of inertia I, and its speed range is from a minimum $\omega_{r\text{-min}}$ to a maximum $\omega_{r\text{-max}}$ rotating speed, so the maximum available kinetic energy, $Ec\text{-}max$ is:

$$E_{c-max} = \frac{1}{2} I(\omega_{r-max}^2 - \omega_{r-min}^2) \tag{3}$$

As Figure 2 shows, the FESS includes two electronic power converters that share a common DC-link built with a bank of capacitors to connect the ASM to the isolated grid. The machine-side converter of Figure 2 controls the ASM in order to keep a constant DC-voltage in the capacitor bank. Therefore, if the capacitor bank voltage rises/falls, this converter discharges/charges the capacitor bank by commanding motor/brake torque to the ASM, so that the flywheel accelerates/decelerates. The rated DC-link voltage and capacitor bank capacitance are 800 V and 4.7 mF, respectively. The grid-side converter of Figure 2 connects the constant voltage DC-link to the isolated grid and is controlled to exchange the necessary active and reactive powers with the grid by setting the corresponding current references. In the present application, this converter works with the power factor equal to unity, so its reactive power reference is zero. As Figure 2 shows, the grid-side converter uses an inductance L-filter to limit the harmonic injection and an isolation transformer is used for coupling with the isolated grid.

The ASM dynamic control is achieved by using Field Oriented Control (FOC). The FOC decouples the control of flux and torque of the ASM [30] and uses a dq-reference frame where the d axis is aligned with the rotor flux vector Ψ_r. The rotor flux is:

$$\Psi_r = L_m i_{mr} \tag{4}$$

where i_{mr} and L_m are the magnetizing current and inductance respectively. The relation between i_{mr} and the stator direct current i_{sd} is given by:

$$T_r \frac{di_{mr}}{dt} + i_{mr} = i_{sd} \tag{5}$$

where T_r is the rotor time constant and $T_r = L_r/R_r$, where R_r and L_r are the rotor resistance and inductance respectively. Therefore, i_{sd} produces the rotor flux Ψ_r, but with slow dynamics due to the high value of T_r. The ASM electromagnetic torque T_{el} is proportional to the product of the rotor flux and the stator quadrature current i_{sq} [31]:

$$T_{el} = \frac{3}{2} \frac{L_m}{L_r} L_m i_{mr} i_{sq} \tag{6}$$

where 3/2 constant considers the 2–3 axes scaling. The FOC keeps the rotor flux constant ($i_{mr} = i_{sd} =$ constant) at its optimal value and the required T_{el} is obtained by setting the corresponding quadrature current i_{sq} according to Equation (6). If all the ASM losses are neglected (stator and rotor resistance and stator iron losses), the exchanged ASM active power can be approximated to the product of the electromagnetic torque T_{el} and the ASM rotor speed ω_r. Also neglecting the FESS double power converter losses, the exchanged FESS active power P_S is approximately:

$$P_S \approx T_{el}\, \omega_r \tag{7}$$

The high flywheel inertia allows considering ω_r speed constant when it is compared to the electric dynamics. Hence, the FESS exchanged power is controlled by setting the needed i_{sq}.

The LS-FESS was sized to store a maximum available energy of 18,000 kJ, to supply the 150 kW FESS rated power ($P_{S\text{-}NOM}$) during 2 min and with the flywheel operating speed range $\omega_{r\text{-}min}$–$\omega_{r\text{-}max}$ within 1500–3300 rpm [8], so Equation (3) gives 380 kg·m^2 for the flywheel moment of inertia I. The selected ASM is a standard 50 Hz one, with a single pole pair and 300 kW of rated power [8]. The model of the 300 kW ASM-FESS uses the block included in the SimPowerSystems blockset for Simulink [16] and its electrical model is a fourth-order one. The flywheel 380 kg m^2 inertia is included in the inertia parameter of the FESS-ASM SimPowerSystems model to simulate the flywheel.

3. The WTG Power Excess Situation and the DL and FESS-off Case Simulation

In the WD mode of the Figure 1 WDPS, the active power produced by the WTG can exceed the load consumption ($P_T > P_L$). If there is no ESS and no DL in the WDPS to consume additional power, the DE should consume the active excess power to balance active powers in the system and thus to control system frequency. However, the DE speed governor cannot command the DE to consume power. In the extreme case of no fuel injection into the DE cylinders, the DE power will be negative ($P_{DE} < 0$) and will consist of losses from compression in the cylinders, shaft, etc., but the DE losses are not controllable. Therefore, if the WTG power excess surpasses the DE losses, the system frequency will rise without control. This frequency increasing is better observed by analyzing the equation which relates the active powers of the DE (P_{DE}), WTG (P_T) and load (P_L) with the DG speed (ω)/frequency (f):

$$P_{DE} + P_T - P_L = J\omega\frac{d\omega}{dt} \tag{8}$$

In Equation (8), the power exchanged by the generators (P_{DE} and P_T) are positive when produced and the consumer load power P_L positive when consumed, J is the DG moment of inertia and the losses are neglected. Since the left side of Equation (8) is positive in the WTG power excess case and J and ω are positive, this implies $d\omega/dt > 0$ and a continuous uncontrolled frequency increase.

All DGs have some shut-down alarms, that is, alarms that, when activated, trip the SM circuit breaker (CB) (I_G in Figure 1) and shut down the DE by cutting the fuel supply to the DE cylinders. Among these alarms is the overspeed alarm, which avoids the uncontrolled acceleration of the DG and therefore excessive centrifugal forces that could damage the rotating parts of the DG. The overspeed alarm is activated when the DG speed is greater than the overspeed setpoint, which is normally 1.1 the DG rated speed. Additionally, all DG have some SM-CB trip alarms, among them the reverse power alarm. This alarm is activated when the DG is connected in parallel with other generators and the DG output power is negative (the SM is behaving as a motor) during a certain time interval. During a DG reverse power activation, the system frequency is supported in the rated value by the other supplying generators.

When there is a WTG power excess ($P_T > P_L$), the DG output power is negative and at the same time the system frequency increases, so that this situation could be confused with an overspeed alarm if the overspeed set point is reached and therefore the protection control will shut down the DG. The WTG power excess case could also be confused with the reverse power alarm if the conditions to active this trip alarm occur. In both cases, the SM-CB will be open and there will not be supply for the consumer load. However, to prevent the discontinuation in the power supply in the WTG power excess case, the solution is to open the WTG CB (I_T in Figure 1) and allow the DG to continue supplying the loads.

The above solution is shown in Figures 3–5, where the Figure 2 WDPS, with both the DL and FESS turned off, is simulated from an initial state where the DG produces 25 kW, the WTG produces 50 kW for a 7 m/s wind speed and the load consumes 75 kW. Figures 3 and 4 show the system frequency and RMS voltage both in pu. Figure 5 shows the active powers in kW for the WTG, DG and load and these active powers are plotted positive/negative when produced/consumed, so that the active powers sum is null at steady state. At $t = 0.2$ s a wind speed step of +2 m/s is applied and consequently the WTG produced power increases. Soon, the WTG power surpasses the load consumption, the DG power becomes negative, the system frequency increases permanently and the WDPS becomes unstable. As Figure 3 shows, when the frequency reaches 1.08 pu (before the 1.1 pu overspeed alarm set point) at $t = 2.935$ s, the control system orders a trip of the WTG 3-phase breaker (3PB) of Figure 2, disconnecting the WTG to protect the WDPS and ensure the power supply continuity. Figure 4 shows increasing voltage oscillations before $t = 2.935$, which corresponds to an unstable system, and presents a strong peak after the tripping of the WTG-3PB. After the WTG disconnection, Figure 3 shows that the system frequency recovers the rated value due to the action of the DE speed governor and Figure 5 shows that in the steady state the DG supplies all the active power (75 kW) to the load.

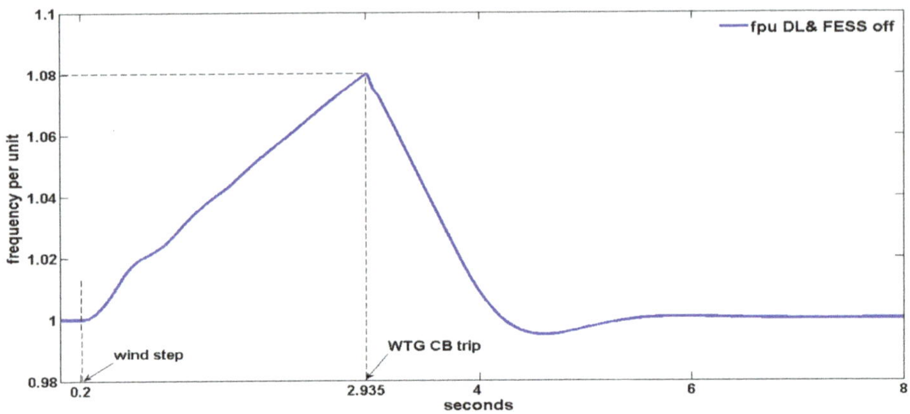

Figure 3. WDPS frequency per unit for the DL and Flywheel Energy Storage System (FESS)-off case.

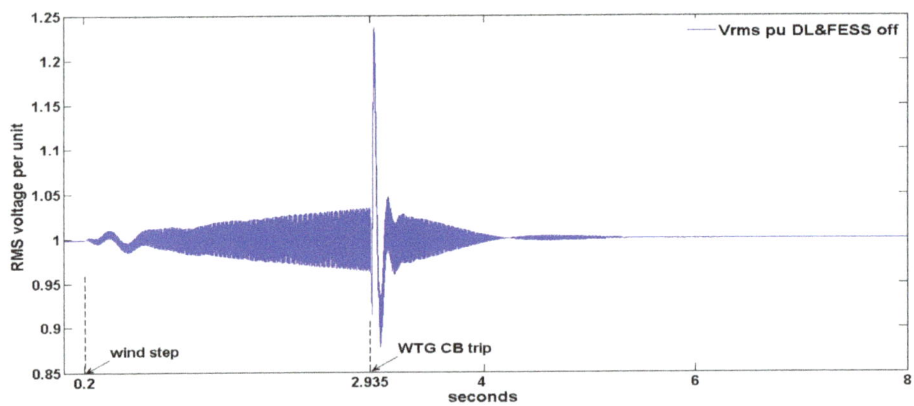

Figure 4. Root Mean Square (RMS) Voltage per unit for the DL and FESS-off case.

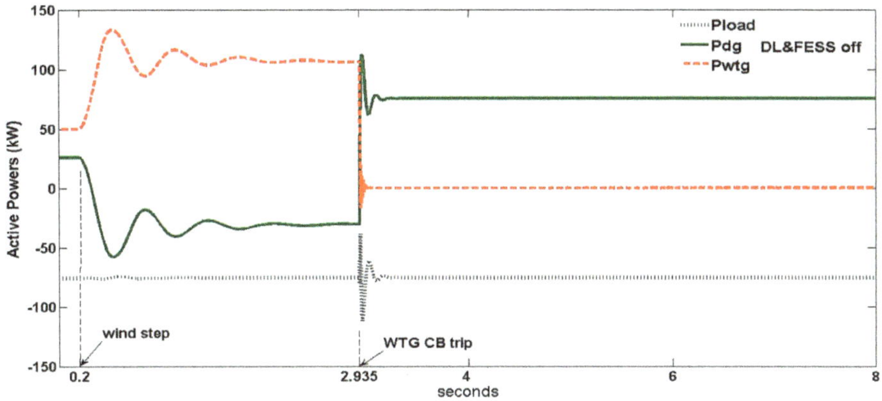

Figure 5. Wind turbine generator (WTG), diesel generator (DG) and load active powers (kW) for the DL and FESS-off case.

4. The WTG Power Excess Situation in the Only-DL/Only-FESS Cases

4.1. The DL and FESS Control

To avoid the DE reverse power and the WDPS frequency increasing that causes the WTG power excess, the WDPS control must command the FESS or DL or both to consume controlled power so that the required DG power to balance the WDPS active power is positive. With a positive DE power, the set speed governor + DE can regulate the WDPS frequency.

In the DL-only case, Equation (8) in steady state ($d\omega/dt = 0$) converts into:

$$P_{DE} = P_L + P_{DL} - P_T \tag{9}$$

where P_{DL} is the DL consumed power and using for P_{DL} the same consumer load sign criteria. As the DL rated power (357 kW) is greater than the WTG rated power (275 kW), Equation (9) indicates that it is possible, by controlling the DL consumed power, to obtain a positive P_{DE} in steady state, even with null load ($P_L = 0$). In this article, the DL dumped power follows the P_{INV} output shown in Figure 6 ($P_{D\text{-}REF} = P_{INV}$). The Figure 6 control is an integral one, and its aim is to keep the DG active power (P_{SM}) in the reference range 15–21 kW (5–7% $P_{DG\text{-}NOM}$) in steady state when a WTG active power excess exists. The control output P_{INV} increases when the DG power P_{SM} is less than 15 kW and decreases its value when P_{SM} is greater than 21 kW. In Figure 6, the integral control is limited to the DL power limits [0, $P_{D\text{-}NOM}$]. The integral constant K_I of Figure 6 was tuned taking into account the used DL discrete nature. To follow P_{INV}, the DL control must order the DL resistors to switch on and off, and this switching produces system voltage variations, as the following simulations graphics show. Therefore, excessive switching must be avoided.

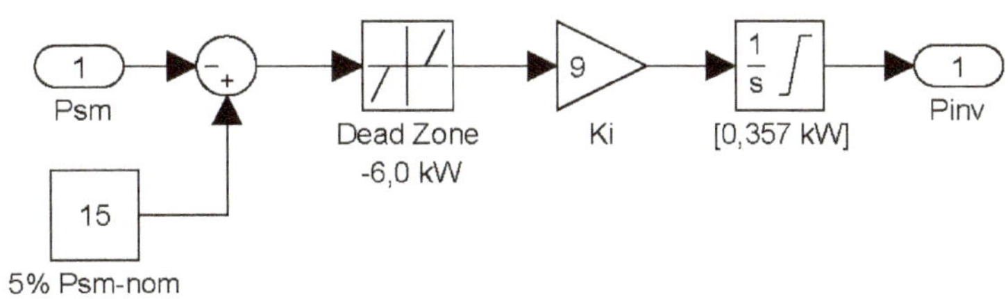

Figure 6. DL/FESS integral control schematic.

In the only-FESS case, Equation (9) also applies if P_{DL} is substituted by the power exchanged by the FESS P_S. As the FESS rated power is 150 kW, the FESS cannot guarantee a positive P_{DE} in steady state if the WTG power excess surpasses 150 kW. In the only-FESS case, the reference power $P_{S\text{-}REF}$ to be consumed by the FESS is given by the following equation:

$$P_{S-REF} = K_p e_f + K_D \frac{de_f}{dt} + P_{INV} \tag{10}$$

where a PD regulator with the frequency error e_f as input ($e_f = f - f_{NOM}$, where f/f_{NOM} are the current/rated WDPS frequency) is added to the formerly explained term P_{INV}. K_P and K_D are the PD proportional and derivate constants. The integral control limits of P_{INV} are [0, $P_{S\text{-}NOM}$] in this only-FESS case. The PD regulator supports frequency regulation, improving the transients of the WDPS and it is compatible with the PID regulator inside the DE speed governor. K_P and K_D were adjusted to moderate the system frequency over/under shooting. Equation (10) makes use of the fast-acting power electronic converter with PWM unlike the DL, which uses the zero-crossing connection of the resistors and this results in slower actuation. Moreover, the DL power is in discrete steps and it may lead to excessive voltage variations as explained previously.

4.2. The Simulation Tests for the Only-DL and Only-FESS Cases

The WDPS initial state for this section's simulations is the same as the previous Section 3. The simulation results are presented in Figures 7–11. The plots of Figures 7 and 8 show the system frequency and RMS voltage in per unit. Graphs that show the active powers in kW are Figure 9 for the DG and WTG and Figure 10 for DL, consumer load and FESS. The same sign criterion of Section 3 is applied to the active powers. The WDPS response is plotted with dot line for the only-DL case and in solid line for the only-FESS case.

As in the previous section, a +2 m/s wind speed step is applied at t = 0.2 s, and the WTG produced power increases. The DG decreases its power commanded by the speed governor, and when it is under the minimum 15 kW level, the Figure 6 integral control commands the DL/FESS to start increasing its power consumption. The DL/FESS increases its power so that the DG power keeps positive and, therefore, the set speed governor +DE can regulate the system frequency. In the steady state, the WTG produces 143 kW, the load consumes 75 kW, the DG produced power is 18.5/19 kW for the DL/FESS cases (a value inside the 15–21 kW set interval) and the DL/ FESS consumes 87.5/87 kW (minimum DG load plus WTG power excess). The initial active power excess leads the frequency to increase and, later on, the frequency undershoots. The fpu minimum–maximum are 0.999–1.0083 pu and 0.9989–1.0042 pu for the DL and FESS cases, respectively. The fpu oscillations are higher in the DL case than in the FESS case. The voltage also oscillates, its minimum–maximum are 0.9859–1.0116 and 0.9944–1.0042 pu for the DL and FESS cases, respectively. The voltage oscillations are bigger in the DL case. The FESS case also obtains less oscillation (better relative stability) in the active powers of the WTG and DG. This a consequence of using fast PWM with high resolution in the FESS converter. In both cases, the frequency maximum is much less than 1.08 pu, so the Figure 2 WTG-3PB is not tripped and the voltage variations are also much smaller than the ones shown in the DL and FESS-off case.

At t = 8 s, the load is increased by 100 kW as Figure 10 shows. The speed governor orders the DG to increase its power production above 21 kW, and the integral control of Figure 6 commands DL/FESS to decrease its consumed power. In DL case, the decreasing consumption of power stops when its power reaches zero and remains at that value, so that the WDPS frequency is regulated solely by the DE speed governor. In the FESS case, the FESS power has small positive peaks due to the PD regulator, which means that FESS supplies power during a brief interval until it reaches zero after the transient, so the FESS briefly supports the DE speed governor. In steady state, the DG supplies 32 kW, the WTG remains supplying the same 143 kW (for the 9 m/s wind speed), the load consumes 175 kW, and the DL/FESS consumption is 0 kW. During the +100 kW transient, the frequency decreases due to the deficit of active power, with a minimum–maximum of 0.9943–1.0014 and 0.9963–1.0002 pu for the DL and FESS cases respectively, being the frequency transient better in the FESS case. The WTG positive peak after the +100 kW of Figure 9 is due to an instantaneous increase in the SCIG slip as the system frequency falls faster than the WTG-SCIG speed. This WTG positive peak counteracts partly the active power shortage, so the WTG-SCIG provides damping to the system. The voltage minimum–maximum are 0.9754–1.0157 and 0.9874–1.011 pu for the DL and FESS cases respectively, so the FESS frequency support also results in a better voltage transient. The FESS case also provides less oscillatory waveforms and, therefore, better relative stability for the WTG and DG active powers.

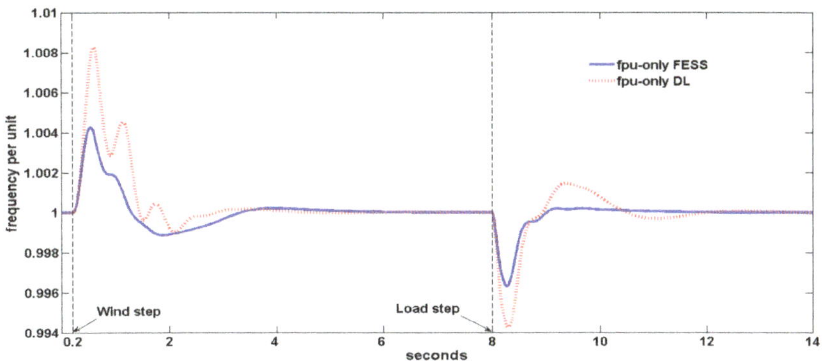

Figure 7. System frequency per unit for the only-FESS and only-DL cases.

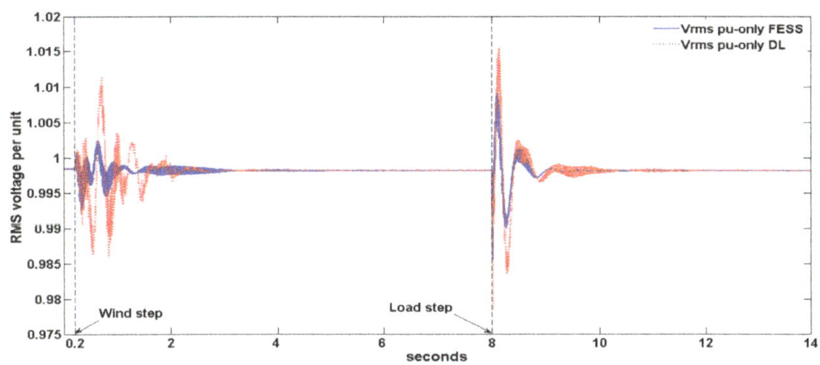

Figure 8. RMS Voltage per unit for the only-FESS and only-DL cases.

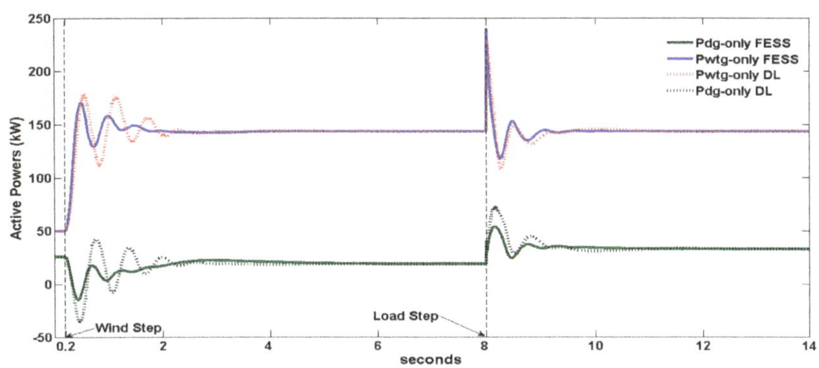

Figure 9. WTG and DG active powers (kW) for the only-FESS and only-DL cases.

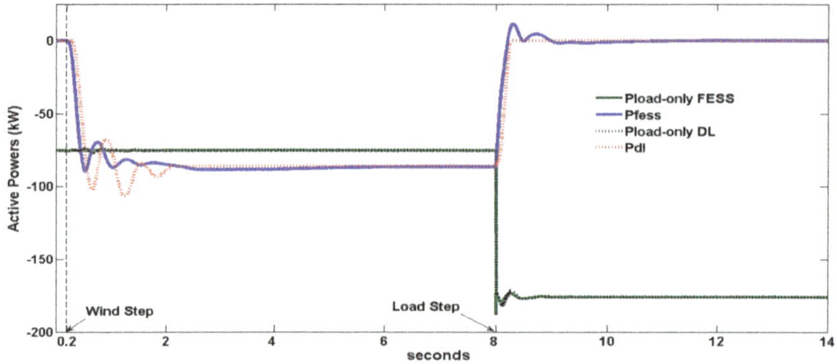

Figure 10. DL, FESS and load active powers (kW) for the only-FESS and only-DL cases.

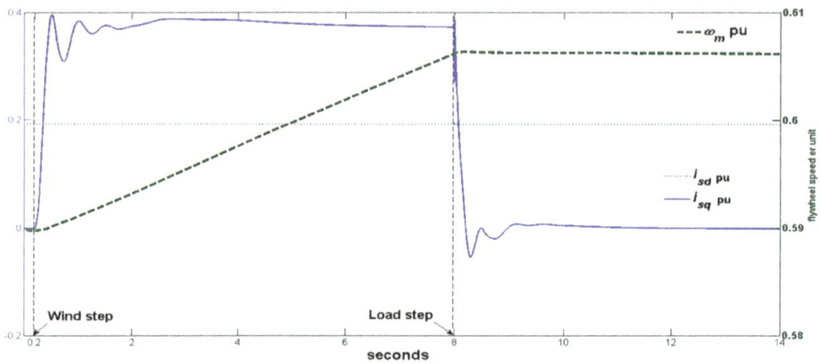

Figure 11. Per unit asynchronous machine (ASM) speed and direct and quadrature currents.

Table 1 summarizes the voltage and frequency variations in percentage for the presented simulations in the three considered cases.

Table 1. Frequency and voltage variations in percentage in the three considered subcases.

Event/Case		DL and FESS-off	Only DL	Only FESS
wind step	%Δf	0, +8	−0.1, +0.83	−0.11, +0.42
	%Δv	−12, +23	−1.41, +1.16	−0.56, +0.42
WTG CB trip		In t = 2.935 s	Not necessary	Not necessary
load step	%Δf	No apply	−0.57, +0.14	−0.37, +0.02
	%Δv		−2.46, +1.57	−1.26, 1.1

4.3. The Flywheel Variables

The direct i_{sd} and quadrature i_{sq} stator currents of the ASM, relative to the 360 A ASM base current and the flywheel shaft speed, relative to the 3000 r.p.m ASM base speed are shown for the FESS test in Figure 11. The direct current creates the ASM magnetic flux, and its sign is always positive. The quadrature current sets the ASM torque and its sign is positive/negative for motor/brake torque (sign criteria usually employed with servos), increasing/decreasing the flywheel speed, and with the FESS consuming/supplying power from/to the isolated grid. There is no PWM current ripple in quadrature and direct currents, since the FESS power converters use average models for faster simulation. The ASM (flywheel) speed indicates how the FESS changes its stored energy, as FESS SOC is proportional to the square of the flywheel speed. Figure 11 left scale is for currents pu and right scale is for speed pu.

Following the ASM-FOC, the direct current value is +0.1923 pu, which is the optimum magnetizing current for maximum power factor at rated power. As explained in Section 2.4, the FESS exchanged power is controlled by adjusting the ASM quadrature current as the ASM-flywheel speed can be considered fairly constant, due to the flywheel high inertia and the short length of the simulations. The initial FESS power and quadrature current are null. After the wind speed step, the 87 kW FESS consumed power is the sum of the WTG power excess (143–75 kW) and the minimum loading of the DG (19 kW), the quadrature current maximum is 0.3951 pu at t = 0.5354 s, and its steady state value is 0.3724 pu. During the +100 kW load step transient, the quadrature current has small negative peaks, which corresponds to the FESS supplying power. After the transient, the FESS power is null, and so the quadrature current returns to null value.

Initially, the FESS speed is 0.5898 pu. After the wind speed step, the FESS speed increases almost linearly as the FESS absorbed power is constant after the transient. After the load step transient, the FESS power is null and the FESS speed remains constant in 0.6061 pu, as FESS losses are not considered.

5. Conclusions

An isolated WDPS with a DL and a LS-FESS was modelled using the Matlab–Simulink framework in Section 2. Section 3 presented, analysed (using Equation (8)) and simulated the WTG power excess scenario in the DL and FESS-off case giving a solution which guarantees power supply continuity with a practical engineering view. Section 4.1 analysed (using Equation (9)) how to guarantee the WDPS absolute stability by using the DL/FESS, and Equation (10) made use of the power supply capability of the FESS to additionally perform frequency support. Section 4.2 presented the simulations for the only-DL and only-FESS cases, showing that both DL and FESS cases prevent the DE reverse power and system frequency increase and therefore it is not necessary to trip the WTG CB unlike the DL and FESS-off case. Table 1 summarized that both DL and FESS cases improve the WDPS power quality by avoiding the strong frequency and voltage perturbations existing in the DL and FESS-off case. Additionally, Figures 7–10 and Table 1 show that the frequency support of the only-FESS case results in lower variations and oscillations in the WDPS frequency and voltage amplitude than the only-DL case. The reason for the better performance of the FESS is the use of its fast action PWM converter with high bandwidth and high resolution instead of the connection of the discrete value resistors at the zero crossing performed by the DL with limited bandwidth.

Further research will include simplifying the dump load by making use of the higher energy density of the last generation FESS [32].

Author Contributions: Conceptualization, R.S.; methodology, R.S., R.P.-A.; software, R.P.-A.; validation, R.S.; investigation, R.S., R.P.-A.; writing—original draft preparation, R.S.; writing—review and editing, R.S., R.P.-A.; visualization, R.S.; supervision, R.S.; project administration, R.S.; All authors have read and agreed to the published version of the manuscript.

Funding: This research received no external funding

Conflicts of Interest: The authors declare no conflict of interest.

List of Abbreviations

ASM: asynchronous machine
CB: Circuit Breaker
DE: Diesel Engine
ESS: Energy storage system; BESS: battery ESS
SCIG: squirrel cage Induction Generator
SM: synchronous machine
WDPS operation modes: Diesel Only (DO), Wind Diesel (WD), Wind Only (WO)
WT: Wind Turbine

Appendix A. Wind Diesel Power System (WDPS) Parameters

WDPS rated frequency and voltage = 60 Hz and 440 V
Diesel Generator (DG) rated power and inertia constant = 300 kVA and 1.75 s.
Wind Turbine Generator (WTG) rated power and inertia constant = 275 kW and 2 s.
Dump Load (DL) rated power = 357 kW
DL integral control $K_I = 9\ s^{-1}$
Flywheel Energy Storage System (FESS); LS-FESS: low speed FESS
FESS rated power, $P_{S\text{-}MOM}$ = 150 kW
FESS maximum available energy = 18,000 kJ
FESS Direct Current (DC) -link voltage = 800 V
FESS capacitor bank capacitance = 4.7 mF
FESS inductance (L) filter = 2.5 mH
FESS Proportional and Derivative (PD) constants: k_P = 151 kW/Hz, K_D = 7 kWs/Hz

References

1. Bhuvaneswari, G.; Balasubramanian, R. 6—Hybrid wind-diesel energy systems. In *Stand-Alone and Hybrid Wind Energy Systems*; Kaldellis, J.K., Ed.; Woodhead Publishing: Sawston, UK, 2010; pp. 191–215. ISBN 9781845695279.
2. Muljadi, E.; McKenna, H.E. Power quality issues in a hybrid power system. *IEEE Trans. Ind. Appl.* **2002**, *38*, 803–809. [CrossRef]
3. Kassem, A.M.; Abdelaziz, A.Y. Functional Predictive Control for Voltage Stability Improvements of Autonomous Hybrid Wind–Diesel Power System. *Electr. Power Comp. Syst.* **2014**, *42*, 831–844. [CrossRef]
4. Janssen, N.T.; Wies, R.W.; Peterson, R.A. Frequency Regulation by Distributed Secondary Loads on Islanded Wind-Powered Microgrids. *IEEE Trans. Sustain. Energy* **2016**, *7*. [CrossRef]
5. Sebastián, R. Battery energy storage for increasing stability and reliability of an isolated Wind Diesel power system. *IET Renew. Power Gener.* **2017**, *11*, 296–303. [CrossRef]
6. Tarkeshwar, M.; Mukherjee, V. Quasi-oppositional harmony search algorithm and fuzzy logic controller for load frequency stabilisation of an isolated hybrid power system. *IET Gener. Trans. Distrib.* **2015**, *9*, 427–444. [CrossRef]
7. Sebastián, R. Application of a battery energy storage for frequency regulation and peak shaving in a wind diesel power system. *IET Gener. Trans. Distrib.* **2016**, *10*, 764–770. [CrossRef]
8. Sebastián, R.; Peña-Alzola, R. Control and simulation of a flywheel energy storage for a wind diesel power system. *Int. J. Electr. Power Energy Syst.* **2015**, *64*, 1049–1056. [CrossRef]
9. Serban, I.; Marinescu, C. Control Strategy of Three-Phase Battery Energy Storage Systems for Frequency Support in Microgrids and with Uninterrupted Supply of Local Loads. *IEEE Trans. Power Electr.* **2014**, *29*, 5010–5020. [CrossRef]
10. Krata, J.; Saha, T.K. Real-Time Coordinated Voltage Support with Battery Energy Storage in a Distribution Grid Equipped with Medium-Scale PV Generation. *IEEE Trans. Smart Grid* **2019**. [CrossRef]
11. Bø, T.I.; Johansen, T.A. Battery Power Smoothing Control in a Marine Electric Power Plant Using Nonlinear Model Predictive Control. *IEEE Trans. Control Syst. Technol.* **2017**, *25*, 1449–1456. [CrossRef]
12. Mendis, N.; Muttaqi, K.M.; Perera, S. Management of Battery-Supercapacitor Hybrid Energy Storage and Synchronous Condenser for Isolated Operation of PMSG Based Variable-Speed Wind Turbine Generating Systems. *IEEE Trans. Smart Grid* **2014**, *5*, 944–953. [CrossRef]
13. Sarasúa, J.; Martínez-Lucas, G.; Platero, C.; Sánchez-Fernández, J. Dual Frequency Regulation in Pumping Mode in a Wind–Hydro Isolated System. *Energies* **2018**, *11*, 2865. [CrossRef]
14. Sarasúa, J.I.; Martínez-Lucas, G.; Lafoz, M. Analysis of alternative frequency control schemes for increasing renewable energy penetration in El Hierro Island power system. *Int. J. Electr. Power Energy Syst.* **2019**, *113*, 807–823. [CrossRef]
15. Sebastián, R.; Alzola, R.P. Flywheel energy storage systems: Review and simulation for an isolated wind power system. *Renew. Sustain. Energy Rev.* **2012**, *16*, 6803–6813. [CrossRef]
16. The MathWorks. SimPowerSystems, Simulink, Block Library Online Documentation. Available online: https://www.mathworks.com/help/physmod/sps/specialized-power-systems.html (accessed on 15 April 2020).

17. Yeager, K.E.; Willis, J.R. Modelling of emergency diesel generators in an 800 Megawatt nuclear power plant. *IEEE Trans. Energy Convers.* **1993**, *8*, 433–441. [CrossRef]
18. Gagnon, R.; Saulnier, B.; Sybille, G.; Giroux, P. Modelling of a Generic High-Penetration No-Storage Wind-Diesel System Using Matlab/Power System Blockset. In Proceedings of the 2002 Global Windpower Conference, Paris, France, 2–5 April 2002.
19. Hansen, A.; Jauch, C.; Sørensen, P.; Iov, F.; Blaabjerg, F. *Dynamic Wind Turbine Models in Power System Simulation Tool DIgSILENT*; Technical Report No. Risø-R-1400(EN); Risø National Laboratory: Roskilde, Denmark, 2003.
20. Li, H.; Chen, Z. Overview of different wind generator systems and their comparisons. *IET Renew. Power Gener.* **2008**, *2*, 123–138. [CrossRef]
21. Kamal, E.; Koutb, M.; Sobaih, A.A.; Abozalam, B. An intelligent maximum power extraction algorithm for hybrid wind-diesel- storage system. *Int. J. Electr. Power Energy Syst.* **2010**, *32*, 170–177. [CrossRef]
22. Haruni, A.M.O.; Gargoom, A.; Haque, M.E.; Negnevitsky, M. Dynamic operation and control of a hybrid wind-diesel stand alone power systems. In Proceedings of the 2010 Twenty-Fifth Annual IEEE Applied Power Electronics Conference and Exposition (APEC), Palm Springs, CA, USA, 21–25 February 2010; pp. 162–169.
23. Lukasievicz, T.; Oliveira, R.; Torrico, C. A Control Approach and Supplementary Controllers for a Stand-Alone System with Predominance of Wind Generation. *Energies* **2018**, *11*, 411. [CrossRef]
24. Tiwari, S.K.; Singh, B.; Goel, P.K. Control of wind diesel hybrid system with bess for optimal operation. In Proceedings of the 2016 IEEE 7th Power India International Conference (PIICON), Bikaner, India, 25–27 November 2016; pp. 1–6.
25. Amenedo, J.L.R.; Diaz, J.C.B.; Gomez, S.A. *Sistemas Eólicos de Producción de Energía Eléctrica*; Editorial Rueda, S.L.: Madrid, Spain, 2003; ISBN 9788472071391.
26. Margaris, I.D.; Papathanassiou, S.A.; Hatziargyriou, N.D.; Hansen, A.D.; Sorensen, P. Frequency Control in Autonomous Power Systems with High Wind Power Penetration. *IEEE Trans. Sustain. Energy* **2012**, *3*, 189–199. [CrossRef]
27. Knudsen, H.; Nielsen, J.N. Introduction to the modeling of wind turbines. In *Wind Power in Power Systems*; Chapter, 24, Ackermann, T., Eds.; Wiley: Chichester, UK, 2005; pp. 525–554.
28. Sebastian, R. Simulation of the transition from wind only mode to wind diesel mode in a no-storage Wind Diesel System. *Latinoam. Trans. IEEE* **2009**, *7*, 539–544. [CrossRef]
29. Hamsic, N.; Schmelter, A.; Mohd, A.; Ortjohann, E.; Schultze, E.; Tuckey, A.; Zimmermann, J. Stabilising the grid voltage and frequency in isolated power systems using a flywheel energy storage system. In Proceedings of the Great Wall World Renewable Energy Forum (GWREF), Beijing, China, 24–27 October 2006.
30. Boldea, I.; Nasar, S.A. *Vector Control of AC Drives*; CRC Press: Boca Raton, Florida, USA, 1992.
31. Cardenas, R.; Pena, R.; Asher, G.M.; Clare, J.; Blasco-Gimenez, R. Control strategies for power smoothing using a flywheel driven by a sensorless vector-controlled induction machine operating in a wide speed range. *IEEE Trans. Ind. Electr.* **2004**, *51*, 603–614. [CrossRef]
32. Lee, B.; Pina, F.; Hope, L.T.; Oglesby, R. *Low-Cost Flywheel Energy Storage Demonstration*; Technical Report No. CEC-500-2015-089; Amber Kinetics: Union City, CA, USA, 2015.

energies

MDPI

Review

Review on Dynamic Simulation of Wind Diesel Isolated Microgrids

Rafael Sebastián

Department of Electrical, Electronic and Control Engineering (DIEEC), Universidad Nacional de Educación a Distancia (UNED), 28040 Madrid, Spain; rsebastian@ieec.uned.es; Tel.: +34-913987624

Abstract: Wind diesel isolated microgrids (WDIMs) combine wind turbine generators (WTGs) with diesel generators (DGs) to supply electricity to remote consumers. WDIMs are low-inertia isolated power systems where large system frequency and voltage variations occur. WDIM dynamic modeling allows short-term simulations to be performed and detailed electrical variable transients to be obtained so that the WDIM power quality and stability can be tested. This paper presents a literature review about WDIM dynamic simulation. The review classifies articles according to factors such as the different WDIM operation modes (diesel only, wind–diesel and wind only) simulated, the types of WTGs used in the WDIM (constant- and variable-speed types), or the use of different short-term energy storage technologies (batteries, ultracapacitors, flywheels) to improve the WDIM power quality, stability and reliability. Papers about the dynamic simulation of related isolated microgrids are also reviewed. Finally, as an example of WDIM dynamic simulation, a WDIM with one WTG, one DG, load and a discrete dump load (DL) is modeled and simulated. The WDIM response to variations of wind speed and load consumption is shown by graphs of the main electrical variables. The simulations show how the DL is used to improve the WDIM stability and reliability.

Keywords: diesel generator; wind turbine generator; isolated microgrid; power system simulation; power quality; power system stability

Citation: Sebastián, R. Review on Dynamic Simulation of Wind Diesel Isolated Microgrids. *Energies* **2021**, *14*, 1812. https://doi.org/10.3390/en14071812

Academic Editor: Antonino Laudani

Received: 17 February 2021
Accepted: 21 March 2021
Published: 24 March 2021

Publisher's Note: MDPI stays neutral with regard to jurisdictional claims in published maps and institutional affiliations.

1. Introduction

Wind diesel isolated microgrids (WDIMs) are microgrids that combine wind turbine generators (WTGs) with diesel generators (DGs) to supply electricity to remote consumers. WDIMs are in many cases the retrofitting of an existing isolated diesel microgrid with WTGs when there is an available wind resource at the diesel microgrid location. By means of the WTG-supplied power, the DG-demanded power is reduced so that the fuel consumption and the CO_2 emissions are also reduced. Figure 1 shows a WDIM scheme where the main components of a WDIM, namely the wind turbine generator (WTG), diesel generator (DG) and consumer load, can be seen. Only one DG and one WTG are shown for simplicity, but WDIMs can include several DGs and/or WTGs. The DG consists of a diesel engine (DE) driving the rotor of a synchronous machine (SM). The DE converts fuel energy into shaft mechanical energy by the combustion of the fuel in the DE cylinders. The DE speed is controlled by the DE speed governor which actuates on the fuel rate incoming into the DE cylinders to control the DE-produced mechanical power. The DE mechanical power is converted into electrical power by the SM, which also provides the sinusoidal grid voltage waveform of frequency f and amplitude V. The SM output voltage amplitude V is controlled by an automatic voltage regulator by regulating the SM reactive power injected in the microgrid. The WTG consists of a wind turbine (WT) driving the rotor of an electrical generator. The WT converts the wind power into shaft mechanical power. The WT mechanical power is converted into electrical power by the WTG electrical generator. The WTG electrical generator can be connected directly to the grid or through a power converter. The dump load (DL) consists of a resistor bank connected to the grid through power switches or an electronic power converter. The DL behaves as a controlled active power consumer. The energy storage system (ESS) consists of short-term energy storage technology suitable for grid applications [1] (based on flywheels, batteries, ultracapacitors, etc.) connected to the grid through an electronic power converter. The ESS behaves as a controlled active power producer/consumer. The reactive power block supplies reactive power to the isolated grid. It can be a synchronous condenser, a static VAR compensator or integrated into the ESS electronic power converter.

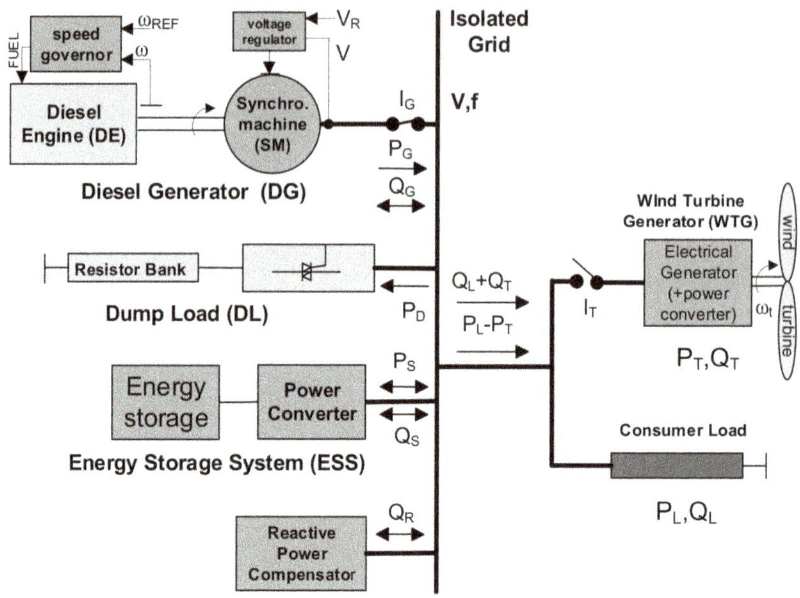

Figure 1. Wind diesel isolated microgrid scheme.

2. WDIM Operation Modes

All WDIMs have two modes of operation: diesel only (DO) and wind–diesel (WD) [2,3]. High-penetration WDIMs are capable of working with the DGs not running working mode known as Wind Only (WO)). The three modes of operation and the conditions to transition among them are shown in Figure 2.

In DO mode, the WDIM behaves as an isolated diesel power plant, and all the consumers' active and reactive power demands are supplied by the diesel generators (WTGs are disconnected in DO mode, I_T = off in Figure 1). The regulation of the frequency is performed by the DE speed governors. The regulation of the voltage is performed by the SM voltage regulators.

The WTGs can supply active power with wind speeds above the WTG cut-in speed and below the cut-off speed. As Figure 2 shows, when the wind speed is inside the previous limits during a predefined T_{WD} time interval to confirm enough wind resource, the WDIM control orders the connection of the WTGs, performing a DO-to-WD mode transition. Conversely, the WDIM control orders the disconnection of the WTGs when the wind speed is outside of the cut-in/out limits (WD-to-DO transition).

In WD mode, both WTGs and DGs supply active power. The regulation of the WDIM voltage and frequency is performed as in DO mode. The DE governor can only control frequency if the DE produces positive power, and a DL is required to guarantee this constraint. The WTG-produced active power P_T can exceed the power consumed by the load P_L ($P_T > P_L$), and if there is no means in the WTG to reduce its power (e.g., by varying the blade pitch), the WDIM control must order the DL to dump the necessary active power P_D to keep the DG power positive P_G and thus to avoid DG reverse power. Therefore, this

DL use increases the WDIM reliability. The ESS can also load the system in WD mode to keep the DG power positive when $P_T > P_L$ [4], with the advantage that the WTG excess power stored in the ESS can later be mostly recovered. ESSs can provide fast reserve power [5] that can be used to reduce the spinning reserve needs in both DO and WD modes. Additionally, for DO/WD modes, ESSs can improve the efficiency of DGs by increasing DG loading (low-loaded DGs have low performance).

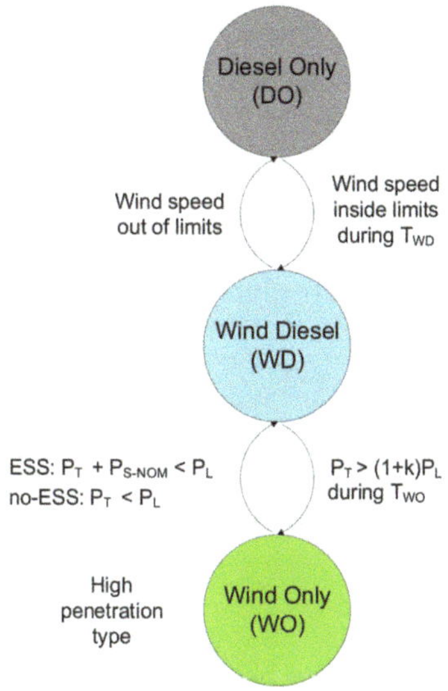

Figure 2. Wind diesel isolated microgrid (WDIM) operation modes and transitions between modes.

As Figure 2 shows, if the WTG-produced power exceeds the load consumption ($P_T > P_L$) (increased by a security factor k) during a predefined T_{WO} time interval, to confirm the WTG active power excess, and the WDIM is a high-penetration one, the WDIM control orders the DGs to stop, performing a WD-to-WO mode transition.

In WO mode, the active power is supplied only by the WTGs. In this mode, as the DGs are not running, no fuel is consumed and the WDIM voltage and frequency control must rely on auxiliary components. In a no-storage WDIM, the condition for WO mode is that the WTG-produced power P_T must exceed the load-consumed power P_L plus the system losses constantly. The active power generated by the WTGs ranges from partly controlled to totally uncontrolled depending on the WTG type. As the consumer load is uncontrolled, the DL is ordered to consume the WTG power excess to match generation with consumption to regulate WDIM frequency. When the WO mode condition is not met ($P_T < P_L$), there will be an active power deficit that will be detected by a system frequency falling, and the WDIM must transition from WO to WD, as Figure 2 shows. The WDIM control will order the DG to start and connect it to the isolated grid. Once the DG is connected, the WDIM will be in WD mode, and the DG will supply power to balance the WDIM active power to restore the frequency to its rated value [6]. If an ESS is included in the WDIM, the WTG power excess can be stored in the ESS or dumped in the DL, or both actions can be taken in a coordinated way. In addition, when the current load is above the WTG power, the ESS can supply power P_S up to its rated power $P_{S\text{-}NOM}$ (temporally until the ESS is discharged). In a WDIM with an ESS, the condition for WO mode is that the WTG-produced power P_T plus $P_{S\text{-}NOM}$ (ESS rated power) must exceed the load consumption P_L (neglecting the losses). As Figure 2 shows, if the previous condition $P_T + P_{S\text{-}NOM} > P_L$ is not satisfied, the WDIM control will start and connect a DG [7], changing from WO to WD mode.

If an SM is running and connected to the grid in WO mode, the SM will generate the microgrid voltage, and the voltage regulation will be performed by the SM voltage regulator as in DO and WD modes. The WDIM active power active unbalances will be translated into frequency variations [8]. If the voltage waveform is generated by a voltage source inverter (VSI) which connects either a WTG or an ESS to the grid, and the VSI works with constant frequency, the WDIM active power unbalances will be translated into voltage variations [9].

3. WDIM Dynamic Simulation Review

WDIM simulations can be of two types: logistic and dynamic. The logistic simulations allow long-term WDIM simulations in terms of energy balance and can be used to test a particular WDIM architecture and for the sizing of the different components of the WDIM. Examples of logistic simulation software are Hybrid2 [10] and Homer [11]. The input data to these programs are the WDIM architecture, the estimated load consumption and the WDIM location wind speed data. These logistic simulation tools produce as output important results such as the fraction of energy that the WTGs supply, the fuel savings achieved, the total DG run time, the number of DG starts and the cost of energy. There are also articles that deal with the generator sizing of a specific WDIM, such as [12], with the aim in this case of optimizing the dynamic power-sharing between the DG and WTG.

WDIM dynamic simulations allow short-term simulations in order to obtain detailed electrical variables waveforms, such as system voltage and frequency and active and reactive power in the different elements, so that the WDIM power quality and stability can be tested. For WDIM dynamic simulations, the most used software are PSCAD [13], which is focused on power system modeling and simulation, and Matlab–Simulink along with the Simscape Electrical [14] library (Simscape Electrical includes power system modeling). The RPM-SIM Simulator [15] has also been used in WDIM dynamic simulation.

In the dynamic simulation of big power systems, the system frequency is assumed to be constant, and this assumption allows the use of low-order dynamic models for the electrical machines in big power system modeling. WDIMs are low-inertia isolated power systems, and therefore, the WDIM frequency can suffer strong variations [7]; thus, according to [16], low-order electrical machine models should not be used in WDIM modeling. Therefore, to obtain precise voltage waveforms in the dynamic simulations, the WDIM modeling must use high-order-type electrical models for the electrical machines included in the WDIM (these high-order models are available on both Simscape Electrical and PSCAD).

This section classifies articles according to several factors, which are dealt with in the following subsections.

3.1. WTGs Included in WDIMs

The three types of generator systems most used in WTGs [17] are shown in Figure 3. The low-cost one is the WTG that equips a squirrel-cage induction generator (SCIG), which has the stator directly connected to the grid (Figure 3a). This combination of a WT driving a SCIG is called a constant-speed WTG, because for generator operation the SCIG speed range is very narrow, typically between 1.01 and 1.02 the synchronous speed [18]. This WT–SCIG set does not allow the variation of its speed to optimize the wind energy capture [17], but it has remarkable features for the remote locations of WDIMs, such as robust construction and simple maintenance [19]. Since the SCIG consumes reactive power, a capacitor bank is added for reactive power compensation. Additionally, the SCIG-produced torque is proportional to the SCIG slip in the WT–SCIG working speed range [20], and due to the SCIG being directly connected to the isolated grid, the WTG inertia increases the system inertia. These two WT–SCIG features provide significant frequency support, moderating system frequency deviations [21]. This WT–SCIG type is used in the WDIM simulations of [22–29].

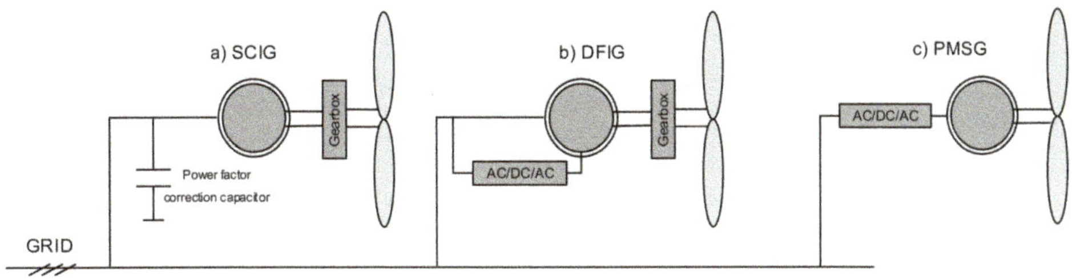

Figure 3. SCIG, DFIG and PMSG Wind Turbine Generators types.

Variable-speed WTGs allow the optimization of the wind energy capture and are mainly of two types [17]: the double-feed induction generator (DFIG) (Figure 3b) type and the permanent magnet synchronous generator (PMSG) type (Figure 3c).

The DFIG type has its stator directly connected to the grid, and its rotor is connected through a slip ring to an AC–DC–AC converter to the grid. This rotor converter controls the rotor frequency and therefore the rotor speed that can be varied in a range of ±30% around the synchronous speed [18]. The rotor converter rated power is only 25–30% of the DFIG rated power, which is an advantage when compared with other variable-speed WTG types. The slip rings need periodic maintenance, which is a drawback for the remote sites of WDIMs. This WT–DFIG type is used in the WDIM simulations of [30–32]. The SCIG and DFIG WTG types need a gearbox to adjust the low WT speed (10–15 rpm) to the high IG rotor speed (for example, 750 rpm for a four-pole-pair IG and 50 Hz grid frequency). The gearbox has friction losses that decrease the WTG efficiency, and it needs periodic maintenance.

The third WTG type considered consists of a WT that directly drives (no gearbox) a PMSG connected to the grid through an AC–DC–AC double converter. The gearbox disadvantages are avoided, but as the SG rotates at the low WT speed, the SG must produce high torque to deliver the rated power; therefore, a larger size of the generator is needed [17]. The double converter makes the WT–SG rotor speed independent from the grid frequency and allows a rotor speed range from 0 to rated one. The double converter rated power has to be equal to the PMSG rated power as all the WTG-produced power has to pass through the converter, so the converter losses are greater than in the WT–DFIG type. In [33–35], the simulated WDIMs include a WT–PMSG type. In the WDIM of [36], the WT drives an electrically excited SG instead of a PMSG. Another interesting feature of the WT–PMSG type is that the DC–AC grid power converter can generate the isolated grid AC voltage and regulate the AC voltage amplitude and frequency, making the WO mode architecture simpler.

Variable-speed operation in WTs has additional advantages such as the reduction in torque peaks in the gearbox and shafts in the DFIG–WTG type and the possibility of using the kinetic energy of the blades to absorb wind power fluctuations, which is an interesting feature in WDIMs. Variable-speed WTGs can also support system frequency if this feature is included in their controllers [21]. Additionally, the AC–DC–AC converter of the variable-speed WTGs allows a battery to be connected to the DC side of the converter [37] so that the battery ESS (BESS) is embedded into the WTG. This embedded BESS can be used to improve the frequency support previously mentioned or to smooth the WTG power output.

3.2. Use of Energy Storage in WDIMs

The aim of the ESSs in a WDIM can be logistic, in order to reduce the start–stop cycles of DGs [38] and/or to improve the WDIM stability and power quality [8]. A significant reduction of the cycling of DGs can be achieved with an ESS with a rated power around the average WDIM load and with a storage time of several minutes, having an acceptable reduction with just 1 min of storage time [38]. In [39], the logistic modeling SW Hybrid2 [10] is used to determine an optimum storage time of 10–15 min for the average WDIM load to reduce the DG cycling. The rated power of WDIMs goes from tens of kW to MW, so this is also the power range of ESSs for WDIMs. In this section, short-term ESSs based on batteries (BESSs), supercapacitors (CESSs) and flywheels (FESSs) are briefly considered as these types comply with the ESS rated power and storage time requirements previously mentioned. Additionally, BESSs, CESSs and FESSs are compared in terms of cycle life, as any ESS in WDIMs must withstand a high number of charging–discharging cycles with variable depth of discharge as the ESS must continuously smooth the wind power, and wind power is of random nature. The structures of the three ESS types considered are presented in Figure 4.

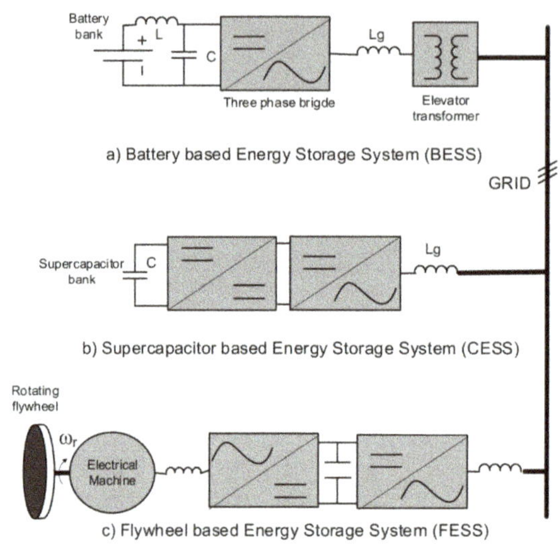

Figure 4. Structures of energy storage systems (ESSs) based on (**a**) batteries (BESSs), (**b**) supercapacitors (CESSs) and (**c**) flywheels (FESSs).

Battery energy storage systems (BESSs) (Figure 4a) consist of a battery bank, a three-phase bidirectional power converter to convert the DC battery voltage into the AC grid voltage and additional filters to smooth the battery current and to limit the grid harmonic injection. The power converter works as an inverter/rectifier to discharge/charge the battery to/from the grid. The modeling of one of these converters working in current-controlled mode can be seen in [40]. In Figure 4a, an elevator transformer matches the battery voltage with the AC grid voltage. This solution is feasible as batteries change voltage little when they discharge, but other solutions instead use a DC–DC buck–boost converter to elevate the battery voltage up to the needed DC input voltage in the inverter [41]. Reference [42] shows different dynamic BESS models. The battery types that have been used in WDIMs are Pb–acid, Ni–Cd, Ni–MH and Li-ion [43]. Reference [44] presents an electrical model for batteries consisting of a variable voltage source in series with a constant resistor. The DC variable voltage source follows the volts–state of charge (SOC) battery discharge curve. Reference [45] improves a previous model by proposing different resistances for charging and discharging, these resistances being a function of SOC. Battery SOC calculation is not simple. Reference [46] proposes battery SOC estimation, integration of the battery current and tracking of battery voltage variations. In [47], the Volterra models are applied to calculate battery SOC. The relationship between the battery number of cycles and the depth of discharge is that of an exponential decrease [48]. The maximum number of cycles for a battery depends on the battery type. WDIM simulations including a BESS in their architecture can be found in [4,19,34,40,49].

Supercapacitor energy storage systems (CESSs) (Figure 4b) consist of a supercapacitor bank, a bidirectional DC–DC converter, a DC–AC three-phase bidirectional power converter and a grid connection filter. Unlike batteries, supercapacitors present significant voltage drop as they discharge, so a DC–DC converter is needed in this case [50] to adapt the supercapacitor DC voltage to the DC voltage needed at the input of the DC–AC grid converter. The simplest electrical model for supercapacitors consists of a series of a capacitor and a resistance along with a parallel resistance which takes account of the autodischarge process. Compared to batteries, supercapacitors have higher specific power (W/kg) and much higher cycle life and allow simpler measurement of the state of charge (by means of the supercapacitor voltage), but they have lower specific energy (Wh/kg) and much higher cost. Reference [51] includes a CESS in its WDIM simulations. By combining the high power density of supercapacitors with the high energy density of batteries, an optimum ESS is achieved [52]. In [35], the WDIM includes a mixed CESS–BESS.

FESSs (Figure 4c) consist of a rotating flywheel, an electrical machine that drives the flywheel and a double AC–DC–AC power converter to connect the electrical machine to the AC grid. Filters are needed to connect the double converter to the grid and to the electrical machine. A capacitor bank is connected to the DC side of the double converter. The energy is stored in the kinetic energy of the flywheel, and the electrical machine acting as a generator/motor converts the kinetic energy into electrical energy and vice versa. The bidirectional electrical machine side converter transforms the DC capacitor bank voltage into AC voltages for the electrical machine. The bidirectional grid side converter transforms the AC grid voltages into the DC voltage of the capacitor bank. The machine converter controls the electrical machine torque in order to maintain a constant DC voltage in the DC-link, so the control of the grid converter works in the same way as a battery converter. Compared to batteries, flywheels have higher cycle life and higher specific power (W/kg) and allow simpler measurement of the state of charge (only flywheel spinning speed is needed), but they have lower specific energy (Wh/kg) and much higher cost.

Reference [53] deals with the modeling of the FESS double power converter. WDIM simulations which include a FESS can be found in [24,25,53,54].

Another type of short-term ESS that has been used in WDIM simulations is superconducting magnetic energy storage (SMES). Reference [55] uses SMES in the presented WDIM. A review on SMES and its potential applications in power systems can be seen in [56].

3.3. Modes of Operation

The literature about WDIM operation mode simulations deals with WDIMs of different architectures. The following papers present WD mode simulations in no-storage WDIMs. Reference [22] deals with a WDIM with one DG and one WTG and shows several simulations considering pitch or stall regulation WTs and WT–SCIG types or WTs driving SGs directly connected to the grid. In [23], the responses of a two-DG–three-WTG WDIM to sudden WTG disconnection and load demand variations are simulated. In the WDIM of [26], modifications to the DE speed governors are proposed to share active power among DGs in order to reduce the total fuel consumption. Reference [57] presents simulations of a WDIM whose fixed-pitch WT drives a dynamic slip-controlled wound rotor induction generator [58]. Reference [57] shows that by varying the rotor WRIG resistance as a function of the grid frequency by means of a variable external resistor, this WT–WRIG provides frequency support to the minigrid. ESSs have also be considered in WD mode simulations. Reference [24] simulates a WDIM with one WTG, one DG, a DL and a FESS based on hydrostatic transmission. In [25], a WDIM includes a high-speed flywheel driven by a switched reluctance electrical machine, and simulations with the WDIM response to WTG power and load increases are presented. Reference [59] models a WDIM with WT–SCIG, DG, BESS, load and power lines connecting all the elements, and it simulates the WDIM in response to a wind gust disturbance.

Several papers deal with simulations in WO mode. In the following papers, an SM provides the isolated grid AC voltage. Reference [29] models an isolated wind power system that consists of a WT–SCIG, SM, load and BESS. The simulation in [29] shows how the BESS is used to regulate frequency by consuming/supplying the WTG active power excess/deficit. Reference [54] deals with a similar architecture, except that the BESS is substituted by a low-speed flywheel driven by an induction electrical machine, and shows how the FESS is used to regulate the isolated system frequency. No-storage WDIMs have also been simulated in WO mode, such as in [6], which shows WDIM frequency regulation by using a DL which consumes the WTG power excess. An architecture similar to that of [6] is presented in [60], but [60] uses a different DL control for frequency regulation. In [61], the regulation of frequency is performed by a BESS and a DL that are coordinated to share the WTG power excess.

In other WO simulations, the grid converter of a WT–PMSG type generates the AC isolated grid voltage, and the frequency regulation is performed by regulating the DC link voltage of the WT–PMSG double power converter. Examples of this case are [34], where a DL and BESS combination regulates the DC link voltage; [35], where a BESS–CESS combination is used for the DC regulation; and [36], only a DL is used to balance the WTG active power generation excess.

The transitions between the different WDIM modes of operation have also been simulated in the literature. Reference [27] simulates a DO-to-WD mode transition by connecting a WT–SCIG to a DG isolated grid. Reference [62] shows a much smoother DO-to-WD transition than [27] as the WT–SCIG is equipped with a soft starter. In [7,49,63], the simulated high-penetration WDIM has a DG with a friction clutch. By means of the clutch, the DE can be locked/disengaged to/from the SM. This clutch-type DG is shown in Figure 5. When the clutch is locked, the WDIM mode is DO–WD; when the clutch is disengaged, the WDIM mode is WO and the SM provides the AC grid voltage. The WO-to-WD transition in the clutch-type DG is faster than in the standard no-clutch DG [64] shown in Figure 1.

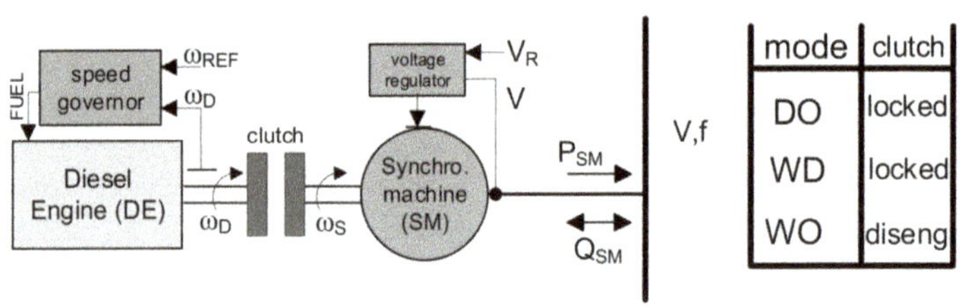

Figure 5. Clutch-type diesel generator (DG).

In the clutch-type DG, the clutch is ordered to engage when the DE and SM have quite similar speeds ($\omega_D \approx \omega_S$ in Figure 5) [7]. The no-clutch DG type shown in Figure 1 needs extra time to perform the synchronization of the SM voltage with the grid voltage before connecting the DG circuit breaker I_G of Figure 1. The WO-to-WD transition simulation in [49] is done by locking the DE to the SM so that the DE produces in WD mode the power supplied previously in WO by the BESS. The WD-to-WO transition simulation of [63] is done by disengaging the DE from the SM when the WTG power generation exceeds the load consumption. Reference [7] simulates the mandatory WO-to-WD transition when power generation is below the power consumption. During this WO-to-WD transition case simulated in [7], there is a frequency falling due to the active power deficit until the clutch is locked. After the locking of the clutch, the DE supplies power and the active power equilibrium can be restored.

3.4. Power Quality and Stability Issues

As previously mentioned, WDIMs are low-inertia isolated power systems, where it is quite difficult to balance power generation with load consumption due to the uncontrolled power generated by the WTGs and consumed by the loads. These unbalances provoke significant system voltage and frequency deviations. Several papers deal with power quality and system stability studies in no-storage WDIMs: [27] studies how the variations of WTG power and load influence WDIM power quality, [65] shows the stabilization of the WDIM voltage by using the combination of a static reactive compensator and the SM voltage regulator, and [26] proposes modifications in the SM voltage regulators to reduce WDIM voltage variations. In [66], the DE speed governor is supported in the frequency regulation by controlling distributed heating loads. In [67], the control of the WTG couples the kinetic energy stored in a WT–DFIG with the rate of change of frequency to emulate inertia, increasing the total inertia of the WDIM and improving the frequency transients.

When a short-term ESS is included in a WDIM, it provides several benefits such as frequency and voltage stabilization and improved stability and reliability [8]. These benefits are more effective in WDIMs than in big power systems which have larger inertia. Reference [51] shows the stabilization of the WDIM frequency by using a capacitive ESS. Reference [68] uses a BESS with a redox flow battery type for frequency stabilization of a WDIM. In [40], the WDIM includes a Ni–MH BESS, and the study compares by simulation the responses for the BESS and no-BESS cases, showing much better transients in voltage and frequency in the BESS case. Similarly, [59] shows by comparing the BESS and no-BESS cases that the DG active and reactive power transients are greatly improved in the BESS case, as the BESS tracks and compensates for the WTG power changes. Reference [69] deals with a WDIM with DL and FESS and shows that the only-FESS case responses present better power quality than the only-DL case when they are used in WD mode to control an active power excess scenario. In [19], the simulated WDIM includes a BESS which performs a peak shaving strategy in order to avoid a DG overload scenario, improving the WDIM reliability. In [4], the WDIM control uses a BESS to increase the power consumption to avoid DG reverse power, so that the DE speed governor can continue controlling system frequency by regulating the DE fuel injection and the WDIM reliability is augmented. Reference [62] also shows how WDIM reliability is increased by ordering a generic ESS to load/supply when needed.

3.5. Related AC Microgrids

WDIMs are remote microgrids [70,71] that operate in an autonomous mode [72], so WDIMs can be related to other isolated microgrid studies. Reference [73] shows a BESS included in an isolated microgrid that provides frequency support and uninterrupted power supply of critical loads; this study is applicable to the ESS in a WDIM in the case of the generators being out of order. Reference [74] deals with a ship with a DG power plant that includes a BESS; the BESS smooths the ship power variation, and this study is applicable to a WDIM working in DO mode. In [75], new DG controllers are proposed to enable the operation of a DG in a microgrid that includes inverter-based sources avoiding circulating reactive power and frequency oscillations caused by standard DG controllers, so this study is applicable to the DGs included in WDIMs. Reference [76] presents graphs of the active and reactive power sharing between a DG and an inverter-based generator in a microgrid when frequency/voltage droop control is used. This study is applicable to a WDIM as ESSs and WT–PMSGs are always connected with an inverter to the grid.

In [77], a BESS is used to counteract the voltage variations caused by renewable power source power fluctuations; [78] presents a standalone microgrid with a BESS–CESS combination and with a fuzzy logic control system to stabilize frequency and voltage; in [79], a load shedding optimal control is carried out to reduce the fuel consumption during the operation time of a DG included in a microgrid with renewable power generation. The utilization of fuel cells to provide controllable active power has also be considered in WDIM and microgrid simulations. In such cases, the microgrid usually includes a water electrolyzer that consumes renewable power excess to produce hydrogen for the fuel cell. Reference [80] models and simulates a WDIM consisting of a DG, four pitch-controlled WTGs, load, a flywheel as short-term ESS and a fuel cell electrolyzer as a long-term ESS, and both ESSs are used to support the WDIM frequency regulation. Reference [81] simulates several cases of a microgrid that includes three WTGs, one DG, a photovoltaic system, two fuel cells and flywheel and battery ESSs. The isolated microgrid in [82] includes wind and photovoltaic generators and fuel cells but does not include a DG. Reference [82] combines a BESS and a CESS to control the frequency of a microgrid. Reference [83] shows an optimal sizing of a BESS used for frequency support in a microgrid with a DG, microturbine, fuel cell and photovoltaic generation.

DLs are also used to balance active power in isolated hydropower systems with no power regulation in the hydro turbine generator (HTG), and thus the DL is used to regulate system frequency [84]. This DL use is the same as that in WO mode in no-ESS WDIMs [6]. Additionally, WO mode simulations are also presented in the wind hydro isolated microgrid of [85], where a simulation of the transition from WO mode to wind–hydro mode (the mode where both the HTG and WTG are supplying) is also shown.

DGs and WTGs are also combined with other renewable power generators, such as in the power plant of El Hierro Island in Spain, which combines wind, hydro and diesel power sources. The diesel-off mode of this wind–hydro–diesel isolated grid has been simulated in several studies. Reference [86] shows how a rotating no-flow Pelton turbine can supply power in less than 10 s when the WTG-produced power is not enough to cover the load demand. In [87], the regulation of the system frequency in WO mode is performed by using the fixed- and variable-speed pumps which belong to the hydropower pumped-storage to balance active power. Reference [88] shows and compares different control schemes for frequency regulation in WO mode in the El Hierro isolated power system.

A broader study about different configurations and control of microgrids can be seen in [89].

4. WDIM Dynamic Modeling and Simulation Example

By using the Matlab–Simulink software, this section models a no-storage WDIM which comprises one 300 kVA DG, one 275 kW fixed-pitch constant-speed WTG, consumer load and a 446 kW discrete DL. After the modeling, the WDIM is simulated, and the system response to wind speed and load variations is shown. The Simulink–SimPowerSystem schematic of the simulated WDIM is presented in Figure 6, and the dynamic modeling of the different components of the WDIM is presented next.

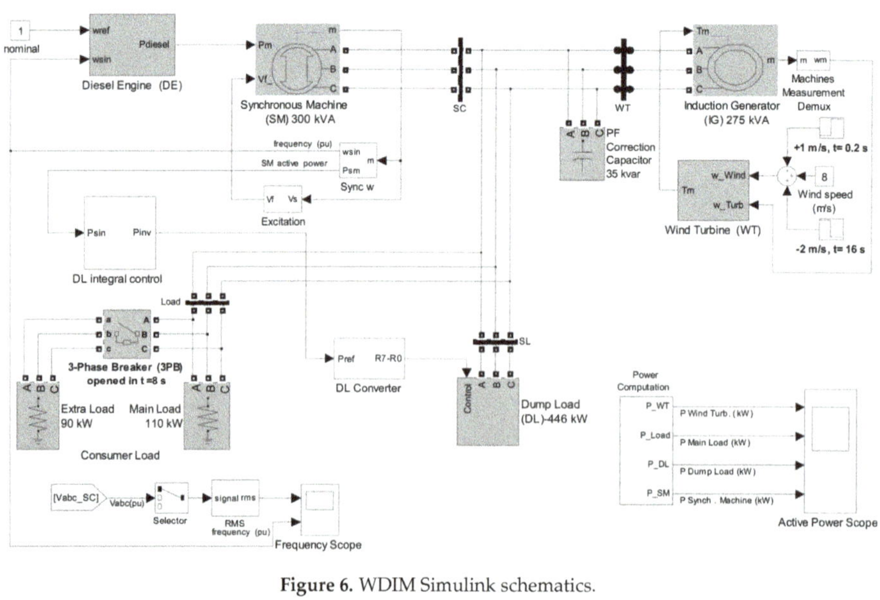

Figure 6. WDIM Simulink schematics.

4.1. Modeling of the WDIM Components

The DG is built with the blocks of Figure 6: diesel engine, 300 kVA SM and voltage regulator. The sixth-order model of the SM and the IEEE type 1 voltage regulator model are available in the Simscape Electrical library [14]. The diesel engine block includes the models of the DE and its speed governor based on [90]. The DE model consists of a gain and a delay, and the speed governor consists of a speed regulator and actuator modeled as a second-order system. The speed regulator is a Proportional-Integral-Derivative (PID) type, and the PID integral part makes the WDIM frequency be the rated one in steady state.

The constant-speed WTG is represented in Figure 6 by the 275 kW induction generator block and the WT block that models the fixed pitch WT. The fourth-order model of the IG is available in the Simscape Electrical library [14]. The WT block is based on the wind turbine power curves [28] and uses a look-up table to calculate the produced mechanical WT shaft power as a function of the wind speed and the WT speed. The WT used has a fixed pitch, so there is no possibility of regulating its produced power, which mainly varies with the cube of the wind speed; therefore, the WT–SCIG generating power is uncontrolled. A 35 kVA capacitor bank for reactive power support of the SCIG is included.

The 446 kW DL consists of eight three-phase resistors (R_0–R_7), each one in series with an electronic switch (S_0–S_7). By closing/opening the switch S_J, the corresponding three-phase resistor R_J is connected/disconnected to/from the isolated grid, so the DL active power consumption can be controlled. The rated powers of resistors R_0–R_7 follow a binary progression and have the values P_0, $2{\cdot}P_0$, $2^2{\cdot}P_0$, … , $2^7{\cdot}P_0$. The DL-consumed power ranges from 0 (all the switches OFF) to $255{\cdot}P_0$ (all the switches ON), and the DL variation is discrete in steps of P_0 kW. For the present application P_0 = 1.75 kW, and so $P_{D\text{-}NOM}$ = 446 kW. The DL model is based on [28]. The DL is used in these simulations to avoid DG reverse power as explained in Section 2. For this aim, an integral action controls the DL-consumed power in such a way that the DG-produced power never goes below 12 kW in steady state. When the $P_T > P_L$ condition happens, the DL is ordered to consume power P_D to allow the DG final power P_G to be within a 12–18 kW minimum power interval (4–6% DG rated power) so that the set DE + speed governor can control WDIM frequency.

4.2. WDIM Simulation Graphics

The simulation graphics show the main WDIM electrical variables: the frequency per unit (fpu) in Figure 7, the RMS voltage per unit in Figure 8 and the active powers of the DG, WTG, DL and consumer load in Figure 9. Figure 9 shows the active powers produced and consumed with positive and negative values, respectively, so the consumer load and DL plots are always below the 0 kW line. At the beginning of the simulation, the WDIM is in steady state, the DG and WTG produce 108 and 92 kW (for a wind speed of 8 m/s), respectively and the load and DL consume 200 and 0 kW, respectively.

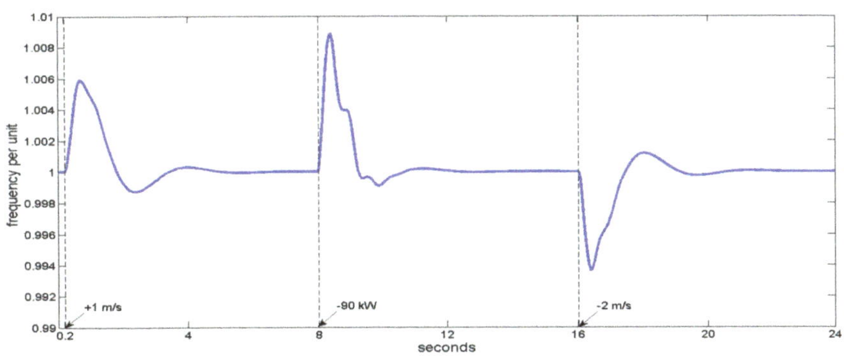

Figure 7. WDIM frequency pu.

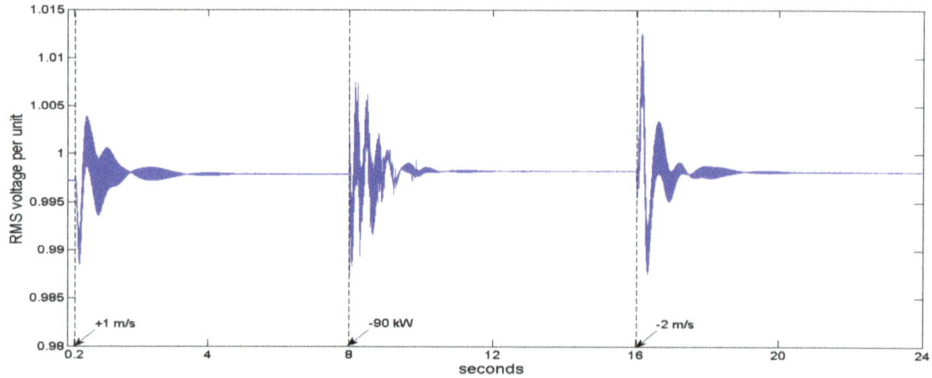

Figure 8. WDIM voltage (RMS) pu.

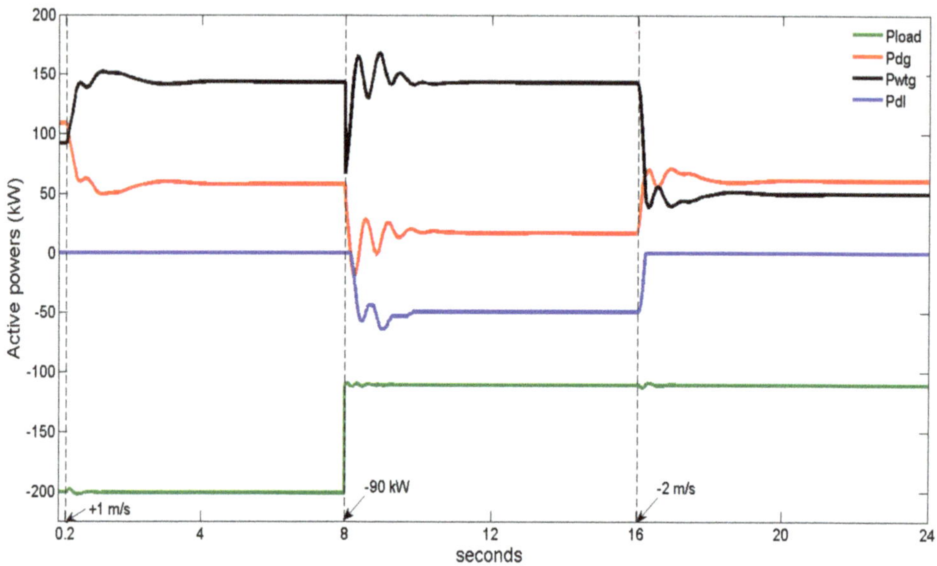

Figure 9. Generated (+)/consumed (-) active powers in the WDIM components.

The wind speed increases at $t = 0.2$ s by 1 m/s to a 9 m/s final value. Figure 9 shows an increasing WTG power from the initial 92 kW to the final 143 KW. The DG consequently decreases its power, and the final DG-produced power is 57 kW. The fpu graph first increases due to the active power excess (1.0059 maximum) and then undershoots (0.9987 minimum), and the voltage pu minimum and maximum are 0.9885 and 1.0039, respectively. The DL is not actuated, and its power is 0 as the DG power remains above 12 kW during the whole 0.2–8 s time interval.

By opening the circuit breaker of Figure 6 at $t = 8$ s, the 90 kW extra load is disconnected, so the consumer load changes to 110 kW, a value below the 143 kW WTG power. The DG reacts by reducing its power, and when its value is below 12 kW, the DL integral control starts ordering power consumption to the DL to prevent the DG reverse power. While the DL increases its consumed power to fit the DG active power in the 12–18 kW interval, both the DG and WTG active powers present oscillations. The initial excess of active power makes the fpu increase (1.0089 maximum), and then the fpu undershoots (0.9991 minimum). The voltage pu minimum and maximum are 0.9873 and 1.0081, respectively. The wind speed remains constant, so the final WTG power is the initial one at $t = 8$ s. In steady state, the DG produces 16.5 kW (a value in the 12–18 range), and the DL consumes 49.5 kW (the DG minimum load plus the WTG active power excess).

At $t = 16$ s, a negative wind speed step of -2 m/s is applied. The WTG power decreases to 50 KW final value. The DG reacts and increases its power over 18 kW so that the DL decreases its consumed power commanded by its integral control. The DL power reaches and stays at zero as the DG power remains greater than 18 KW. The initial lack of active power makes the fpu decrease (0.9937 minimum), and then it overshoots (1.0012 maximum). The voltage pu minimum and maximum are 0.9875 and 1.0126, respectively. The final active powers for the DG, WTG and DL are 60, 50 and 0 kW, respectively.

5. Conclusions

After a description of the different components that form a WDIM and the different modes of WDIM operation and a brief introduction about WDIM dynamic and logistic simulation, papers about the dynamic simulation of WDIMs have been reviewed. The review has been classified according to several WDIM issues: the WTG types that WDIMs include, the use of different ESS technologies in WDIMs, the simulation of WDIM operation modes (or simulation of transitions between modes) and the WDIM power quality and stability studies. Additionally, a brief review of WDIM-related isolated microgrids has been presented. Finally, as an example of WDIM dynamic simulation, a no-storage WDIM has been first modeled and then simulated using Matlab–Simulink SW. In the presented simulations, the WDIM works in WD mode, and it is shown how a DL increases the WDIM stability and reliability by consuming the WTG power excess.

Funding: This research received no external funding.

Institutional Review Board Statement: Not applicable.

Informed Consent Statement: Not applicable.

Data Availability Statement: Not applicable.

Conflicts of Interest: The author declares no conflict of interest.

Abbreviations

DE	Diesel engine
DFIG	Double-feed induction generator
DG	Diesel generator
DL	Dump load
ESS	Energy storage system
BESS	Battery ESS
FESS	Flywheel ESS
CESS	Supercapacitor ESS
PMSG	Permanent magnet synchronous generator
SCIG	Squirrel-cage induction generator
SM	Synchronous machine
WDIM	Wind diesel isolated microgrid
WDIM operation modes	Diesel only (DO), wind–diesel (WD), wind only (WO)
WT	Wind turbine
WTG	Wind turbine generator

References

1. Vazquez, S.; Lukic, S.M.; Galvan, E.; Franquelo, L.G.; Carrasco, J.M. Energy Storage Systems for Transport and Grid Applications. *IEEE Trans. Ind. Electron.* **2010**, *57*, 3881–3895. [CrossRef]
2. Hunter, R.; Elliot, G. (Eds.) *Wind-Diesel Systems: A Guide to the Technology and Its Implementations*; Cambridge University Press: Cambridge, UK, 1994.

3. Sebastián, R.; García-Loro, F. Review on Wind Diesel Systems Dynamic Simulation. In Proceedings of the IECON 2019—45th Annual Conference of the IEEE Industrial Electronics Society, Lisbon, Portugal, 14–17 October; IEEE: New York, NY, USA, 2019; pp. 2489–2494.
4. Sebastián, R. Reverse power management in a wind diesel system with a battery energy storage. *Int. J. Electr. Power Energy Syst.* **2013**, *44*, 160–167. [CrossRef]
5. Psarros, G.N.; Karamanou, E.G.; Papathanassiou, S.A. Feasibility Analysis of Centralized Storage Facilities in Isolated Grids. *IEEE Trans. Sustain. Energy* **2018**, *9*, 1822–1832. [CrossRef]
6. Fernandez, R.S. Simulation of the transition from Wind only mode to Wind Diesel mode in a no-storage Wind Diesel System. *IEEE Latin Am. Trans.* **2009**, *7*, 539–544. [CrossRef]
7. Sebastián, R. Smooth transition from wind only to wind diesel mode in an autonomous wind diesel system with a battery-based energy storage system. *Renew. Energy* **2008**, *33*, 454–466. [CrossRef]
8. Sebastián, R. Battery energy storage for increasing stability and reliability of an isolated Wind Diesel power system. *IET Renew. Power Gener.* **2017**, *11*, 296–303. [CrossRef]
9. Drouilhet, S. *Power Flow Management in a High Penetration Wind–Diesel Hybrid Power System with a Short-Term Energy Storage*; Wind Power 99: Vermont, VT, USA, 1999.
10. Manwell, J.F.; Rogers, A.; Hayman, G.; Avelar, C.T.; McGowan, J.G.; Abdulwahid, U.; Wu, K. *Hybrid2—A Hybrid System Simulation Model Theory Manual*; Renewable Energy Research Laboratory, Department of Mechanical Engineering, University of Massachusetts: Amherst, MA, USA, 2006.
11. Lambert, T.; Gilman, P.; Lilienthal, P. *Micropower System Modeling with HOMER, Integration of Alternative Sources of Energy*; Farret, F.A., Simões, M.G., Eds.; John Wiley & Sons: Hoboken, NJ, USA, 2005; ISBN 0471712329.
12. Saha, T.K.; Kastha, D. Design Optimization and Dynamic Performance Analysis of a Stand-Alone Hybrid Wind–Diesel Electrical Power Generation System. *IEEE Trans. Energy Convers.* **2010**, *25*, 1209–1217. [CrossRef]
13. PSCAD. Manitoba Hydro International Ltd. Available online: https://www.pscad.com/ (accessed on 26 November 2020).
14. Simscape Electrical™ Libraries. Available online: https://es.mathworks.com/products/simscape-electrical.html (accessed on 26 November 2020).
15. Bialasiewicz, J.T.; Muljadi, E. Analysis of Renewable-Energy Systems Using RPM-SIM Simulator. *IEEE Trans. Ind. Electron.* **2006**, 53. [CrossRef]
16. Krause, P.C.; Wasynczuk, O.; Sudhoff, S.D. *Analysis of Electric Machinery and Drive Systems*, 2nd ed.; Wiley-IEEE Press: Piscataway, NJ, USA, 2002.
17. Li, H.; Chen, Z. Overview of different wind generator systems and their comparisons. *IET Renew. Power Gener.* **2008**, *2*, 123–138. [CrossRef]
18. Hansen, A.; Jauch, C.; Sørensen, P.; Iov, F.; Blaabjerg, F. *Dynamic Wind Turbine Models in Power System Simulation Tool DIgSILENT*; Report Risø-R-1400(EN); Risø National Laboratory: Roskilde, Denmark, 2003. Available online: http://www.risoe.dk/rispubl/VEA/veapdf/ris-r-1400.pdf (accessed on 21 March 2021).
19. Sebastián, R. Application of a battery energy storage for frequency regulation and peak shaving in a wind diesel power system. *IET Gener. Transm. Distrib.* **2016**, *10*, 764–770. [CrossRef]
20. Amenedo, J.L.R.; Diaz, J.C.B.; Gomez, S.A. *Sistemas Eólicos de Producción de Energía Eléctrica*; Rueda: Madrid, Spain, 2003; ISBN 9788472071391.
21. Margaris, I.D.; Papathanassiou, S.A.; Hatziargyriou, N.D.; Hansen, A.D.; Sorensen, P. Frequency Control in Autonomous Power Systems With High Wind Power Penetration. *IEEE Trans. Sustain. Energy* **2012**, *3*, 189–199. [CrossRef]
22. Papathanassiou, S.A.; Papadopoulos, M.P. Dynamic characteristics of autonomous wind–diesel systems. *Renew. Energy* **2001**, *23*, 293–311. [CrossRef]
23. Sedaghat, B.; Jalilvand, A.; Noroozian, R. Design of a multilevel control strategyfor integration of stand-alone wind/diesel system. *Int. J. Electr. Power Energy Syst.* **2012**, *35*, 123–137. [CrossRef]
24. Carrillo, C.; Feijóo, A.; Cidrás, J. Comparative study of flywheel systems in an isolated wind plant. *Renew. Energy* **2008**. [CrossRef]
25. Iglesias, I.J.; Garcia-Tabares, L.; Agudo, A.; Cruz, I.; Arribas, L. Design and simulation of a stand-alone wind-diesel generator with a flywheel energy storage system to supply the required active and reactive power. In Proceedings of the 2000 IEEE 31st Annual Power Electronics Specialists Conference, Conference Proceedings (Cat. No.00CH37018), Galway, Ireland, 18–23 June 2000; Volume 3, pp. 1381–1386.
26. Esmaeilian, H.R.; Fadaeinedjad, R. A Remedy for Mitigating Voltage Fluctuations in Small Remote Wind-Diesel Systems Using Network Theory Concepts. *IEEE Trans. Smart Grid* **2018**, *9*, 4162–4171. [CrossRef]
27. Muljadi, E.; McKenna, H.E. Power quality issues in a hybrid power system. *IEEE Trans. Ind. Appl.* **2002**, *38*, 803–809. [CrossRef]
28. Gagnon, R.; Saulnier, B.; Sybille, G.; Giroux, P. Modelling of a Generic High-Penetration No-Storage Wind-Diesel System Using Matlab/Power System Blockset. In Proceedings of the 2002 Global Windpower Conference, Paris, France, 2–5 April 2002.
29. Sebastián, R.; Quesada, J. Distributed control system for frequency control in a isolated wind system. *Renew. Energy* **2006**, *31*, 285–305. [CrossRef]
30. Peña, R.; Cárdenas, R.; Proboste, J.; Clare, J.; Asher, G. Wind–Diesel Generation Using Doubly Fed Induction Machines. *IEEE Trans. Energy Convers.* **2008**, *23*, 202–214. [CrossRef]

31. Tiwari, S.K.; Singh, B.; Goel, P.K. Control of wind diesel hybrid system with BESS for optimal operation. In Proceedings of the 2016 IEEE 7th Power India International Conference (PIICON), Bikaner, India, 25–27 November 2016; pp. 1–6.
32. Kayikci, M.; Milanovic, J.V. Dynamic Contribution of DFIG-Based Wind Plants to System Frequency Disturbances. *IEEE Trans. Power Syst.* **2009**, *24*, 859–867. [CrossRef]
33. Kamal, E.; Koutb, M.; Sobaih, A.A.; Abozalam, B. An intelligent maximum power extraction algorithm for hybrid wind-diesel-storage system. *Int. J. Electr. Power Energy Syst.* **2010**, *32*, 170–177. [CrossRef]
34. Haruni, A.M.O.; Gargoom, A.; Haque, M.E.; Negnevitsky, M. Dynamic operation and control of a hybrid wind-diesel stand alone power systems. In Proceedings of the 2010 Twenty-Fifth Annual IEEE Applied Power Electronics Conference and Exposition (APEC), Palm Springs, CA, USA, 21–25 February 2010; pp. 162–169.
35. Mendis, N.; Muttaqi, K.M.; Perera, S. Management of Battery-Supercapacitor Hybrid Energy Storage and Synchronous Condenser for Isolated Operation of PMSG Based Variable-Speed Wind Turbine Generating Systems. *IEEE Trans. Smart Grid* **2014**, *5*, 944–953. [CrossRef]
36. Lukasievicz, T.; Oliveira, R.; Torrico, C. A Control Approach and Supplementary Controllers for a Stand-Alone System with Predominance of Wind Generation. *Energies* **2018**, *11*, 411. [CrossRef]
37. Yang, T.C. On embedded energy storage for high penetration of wind power. *Wind Eng.* **2008**, *32*, 223–242. [CrossRef]
38. Beyer, H.G.; Degner, T.; Gabler, H. Operational behaviour of wind diesel systems incorporating short-term storage: An analysis via simulation calculations. *Sol. Energy* **1995**, *54*, 429–439. [CrossRef]
39. Shirazi, M.; Droulihet, S. An analysis of the performance benefits of Short-Term Energy Storage in Wind-Diesel Hybrid Power Systems. In Proceedings of the ASME Wind Energy Symposium, Reno, NV, USA, 6–9 January 1997.
40. Sebastián, R.; Peña Alzola, R. Simulation of an isolated Wind Diesel System with battery energy storage. *Electr. Power Syst. Res.* **2011**, *81*, 677–686. [CrossRef]
41. Rezkallah, M.; Chandra, A. Wind diesel battery hybrid system with power quality improvement for remote communities. In Proceedings of the IEEE Industry Applications Society Annual Meeting (IAS), Orlando, FL, USA, 9–13 October 2011; pp. 1–6.
42. Denmark Adrees, A.; Andami, H.; Milanović, J.V. Comparison of dynamic models of battery energy storage for frequency regulation in power system. In Proceedings of the 2016 18th Mediterranean Electrotechnical Conference (MELECON), Lemesos, Cyprus, 18–20 April 2016; pp. 1–6. [CrossRef]
43. Sebastián, R.; Peña-Alzola, R. Study and simulation of a battery based energy storage system for wind diesel hybrid systems. In Proceedings of the 2012 IEEE International Energy Conference and Exhibition (ENERGYCON), Florence, Italy, 9–12 September; 2012; pp. 563–568.
44. Tremblay, O.; Dessaint, L.-A.; Dekkiche, A.-I. A generic battery model for the dynamic simulation of hybrid electric vehicles. In Proceedings of the Vehicle Power and Propulsion Conference, 2007, VPPC 2007, Arlington, TX, USA, 9–12 September; 2007; pp. 284–289.
45. Li, X.; Hui, D.; Lai, X. Battery energy storage station (BESS)-based smoothing control of photovoltaic (PV) and wind power generation fluctuations. *IEEE Trans. Sustain. Energy* **2013**, *4*, 464–473. [CrossRef]
46. Coleman, M.; Zhu, C.B.; Lee, C.K.; Hurley, W.G. A combined SOC estimation method under varied ambient temperature for a lead-acid battery. In Proceedings of the Twentieth Annual IEEE Applied Power Electronics Conference and Exposition, 2005, APEC 2005, Austin, TX, USA, 6–10 March; 2005; Volume 2, pp. 991–997.
47. Sidorov, D.; Muftahov, I.; Tomin, N.; Karamov, D.; Panasetsky, D.; Dreglea, A.; Foley, A. A Dynamic Analysis of Energy Storage With Renewable and Diesel Generation Using Volterra Equations. *IEEE Trans. Ind. Inform.* **2020**, *16*, 3451–3459. [CrossRef]
48. Bindner, H.; Cronin, T.; Lundsager, P.; Manwell, J.F.; Abdulwahid, U.; Baring-Gould, I. *Lifetime Modelling of Lead Acid Batteries*; Risø National Laboratory Report; Risø National Laboratory: Roskilde, Denmark, 2005; 81p, ISBN 87-550-3441-1.
49. Sebastian, R. Modelling and simulation of a high penetration wind diesel system with battery energy storage. *Int. J. Electr. Power Energy Syst.* **2011**, *33*, 767–774. [CrossRef]
50. Chau, K.T. 21—Pure electric vehicles. In *Alternative Fuels and Advanced Vehicle Technologies for Improved Environmental Performance*; Richard, F., Ed.; Woodhead Publishing: Sawston, UK; pp. 655–684, ISBN 9780857095220.
51. Tarkeshwar, M.; Mukherjee, V. Quasi-oppositional harmony search algorithm and fuzzy logic controller for load frequency stabilisation of an isolated hybrid power system. *IET Gener. Transm. Distrib.* **2015**, *9*, 427–444. [CrossRef]
52. Perelmuter, V.M. *Advanced Simulation of Alternative Energy Simulation with Simulink® and SimPowerSystems™*; CRC Press: Boca Raton, FL, USA, 2020; ISBN 9780367339579.
53. Sebastián, R.; Peña-Alzola, R. Control and simulation of a flywheel energy storage for a wind diesel power system. *Int. J. Electr. Power Energy Syst.* **2015**, *64*, 1049–1056. [CrossRef]
54. Sebastián, R.; Alzola, R.P. Flywheel energy storage systems: Review and simulation for an isolated wind power system. *Renew. Sustain. Energy Rev.* **2012**, *16*, 6803–6813. [CrossRef]
55. Zargar, M.Y.; Mufti MU, D.; Lone, S.A. Adaptive predictive control of a small capacity SMES unit for improved frequency control of a wind-diesel power system. *IET Renew. Power Gener.* **2017**, *11*, 1832–1840. [CrossRef]
56. Ali, M.H.; Wu, B.; Dougal, R.A. An Overview of SMES Applications in Power and Energy Systems. *IEEE Trans. Sustain. Energy* **2010**, *1*, 38–47. [CrossRef]
57. Mipoung, O.D.; Lopes, L.A.C.; Pillay, P. Frequency Support From a Fixed-Pitch Type-2 Wind Turbine in a Diesel Hybrid Mini-Grid. *IEEE Trans. Sustain. Energy* **2014**, *5*, 110–118. [CrossRef]

58. Mahela, O.P.; Shaik, A.G. Comprehensive overview of grid interfaced wind energy generation systems. *Renew. Sustain. Energy Rev.* **2016**, *57*, 260–281. [CrossRef]
59. Tang, X.; Deng, W.; Qi, Z. Investigation of the Dynamic Stability of Microgrid. *IEEE Trans. Power Syst.* **2014**, *29*, 698–706. [CrossRef]
60. Kassem, A.M.; Abdelaziz, A.Y. Optimal control of an autonomous variable speed wind generation system based on a bacterial foraging optimization technique. *Electr. Power Compon. Syst.* **2015**, *43*, 1006–1017. [CrossRef]
61. Sebastián, R.; Quesada, J. Coordinated control of a battery energy storage and a Dump load in an autonomous Wind Power System. In Proceedings of the 2020 IEEE International Conference on Environment and Electrical Engineering and 2020 IEEE Industrial and Commercial Power Systems Europe (EEEIC/I&CPS Europe), Madrid, Spain, 9–12 June 2020; pp. 1–6. [CrossRef]
62. Muljadi, E.; Bialasiewicz, J.T. Hybrid power system with a controlled energy storage. In Proceedings of the Industrial Electronics Conference, IECON'03, Roanoke, VA, USA, 2–6 November 2003; Volume 2, pp. 1296–1301.
63. Sebastián, R.; Alzola, R.P. Effective active power control of a high penetration wind diesel system with a Ni–Cd battery energy storage. *Renew. Energy* **2010**, *35*, 952–965. [CrossRef]
64. Sebastian, R.; Castro, M.; Sancristobal, E.; Yeves, F.; Peire, J.; Quesada, J. Approaching hybrid wind-diesel systems and controller area network. In Proceedings of the IEEE 2002 28th Annual Conference of the Industrial Electronics Society, IECON 02, Sevilla, Spain, 5–8 November 2002; Volume 3, pp. 2300–2305.
65. Kassem, A.M.; Abdelaziz, A.Y. Reactive power control for voltage stability of standalone hybrid wind–diesel power system based on functional model predictive control. *IET Renew. Power Gener.* **2014**, *8*, 887–899. [CrossRef]
66. Janssen, N.T.; Wies, R.W.; Peterson, R.A. Frequency regulation by distributed secondary loads on islanded wind-powered microgrids. *IEEE Trans. Sustain. Energy* **2016**, *7*, 1028–1035. [CrossRef]
67. Zhang, Y.; Melin, A.M.; Djouadi, S.M.; Olama, M.M.; Tomsovic, K. Provision for guaranteed inertial response in diesel-wind systems via model reference control. *IEEE Trans. Power Syst.* **2018**, *33*, 6557–6568. [CrossRef]
68. Kouba NE, Y.; Menaa, M.; Hasni, M.; Boudour, M. A novel optimal frequency control strategy for an isolated wind–diesel hybrid system with energy storage devices. *Wind Eng.* **2016**, *40*, 497–517. [CrossRef]
69. Sebastián, R.; Peña-Alzola, R. Flywheel Energy Storage and Dump Load to Control the Active Power Excess in a Wind Diesel Power System. *Energies* **2020**, *13*, 2029. [CrossRef]
70. Lasseter, R. Microgrids. In Proceedings of the IEEE PES Winter Meeting, New York, NY, USA, 27–31 January 2002; Volume 1, pp. 305–308.
71. Basak, P.; Chowdhury, S.; Dey, S.H.N.; Chowdhury, S.P. A literature review on integration of distributed energy resources in the perspective of control, protection and stability of microgrid. *Renew. Sustain. Energy Rev.* **2012**, *16*, 5545–5556. [CrossRef]
72. Piagi, P.; Lasseter, R.H. Autonomous Control of Microgrids. In Proceedings of the IEEE PES Meeting, Montreal, QC, Canada, 19 June 2006.
73. Serban, I.; Marinescu, C. Control Strategy of Three-Phase Battery Energy Storage Systems for Frequency Support in Microgrids and with Uninterrupted Supply of Local Loads. *IEEE Trans. Power Electron.* **2014**, *29*, 5010–5020. [CrossRef]
74. Bø, T.I.; Johansen, T.A. Battery Power Smoothing Control in a Marine Electric Power Plant Using Nonlinear Model Predictive Control. *IEEE Trans. Control Syst. Technol.* **2017**, *25*, 1449–1456. [CrossRef]
75. Krishnamurthy, S.; Jahns, T.M.; Lasseter, R.H. The operation of diesel gensets in a CERTS microgrid. In Proceedings of the 2008 IEEE Power and Energy Society General Meeting—Conversion and Delivery of Electrical Energy in the 21st Century, Pittsburgh, PA, USA, 20–24 July 2008; pp. 1–8.
76. Tan, Y.; Meegahapola, L.; Muttaqi, K.M. A review of technical challenges in planning and operation of remote area power supply systems. *Renew. Sustain. Energy Rev.* **2014**, *38*, 876–889. [CrossRef]
77. Krata, J.; Saha, T.K. Real-Time Coordinated Voltage Support with Battery Energy Storage in a Distribution Grid Equipped with Medium-Scale PV Generation. *IEEE Trans. Smart Grid* **2018**, *10*, 3486–3497. [CrossRef]
78. Asghar, F.; Talha, M.; Kim, S.H. Fuzzy logic based intelligent frequency and voltage stability control system for standalone microgrid. *Int. Trans. Electr. Energy Syst.* **2018**, *28*, e2510. [CrossRef]
79. Madiba, T.; Bansal, R.C. Optimal Load-Shedding Control of a Microgrid Power System. *Electr. Power Compon. Syst.* **2018**, *46*, 768–787. [CrossRef]
80. Vidyanandan, K.V.; Nilanjan, S. Frequency regulation in a wind–diesel powered microgrid using flywheels and fuel cells. *IET Gener. Transm. Distrib.* **2016**, *10*, 780–788. [CrossRef]
81. Lee, D.; Wang, L. Small-Signal Stability Analysis of an Autonomous Hybrid Renewable Energy Power Generation/Energy Storage System Part I: Time-Domain Simulations. *IEEE Trans. Energy Convers.* **2008**, *23*, 311–320. [CrossRef]
82. Taghizadeh, M.; Mardaneh, M.; Sadeghi, M.S. Frequency control of a new topology in proton exchange membrane fuel cell/wind turbine/photovoltaic/ultra-capacitor/battery energy storage system based isolated networks by a novel intelligent controller. *J. Renew. Sustain. Energy* **2014**, *6*, 053121. [CrossRef]
83. Aghamohammadi, M.R.; Abdolahinia, H. A new approach for optimal sizing of battery energy storage system for primary frequency control of islanded Microgrid. *Int. J. Electr. Power Energy Syst.* **2014**, *54*, 325–333. [CrossRef]
84. Paish, O. Small hydro power: Technology and current status. *Renew. Sustain. Energy Rev.* **2002**, *6*, 537–556. [CrossRef]
85. Sebastián, R.; Nevado, A. Study and Simulation of a Wind Hydro Isolated Microgrid. *Energies* **2020**, *13*, 5937. [CrossRef]

86. Platero, C.A.; Nicolet, C.; Sánchez, J.A.; Kawkabani, B. Increasing wind power penetration in autonomous power systems through no-flow operation of Pelton turbines. *Renew. Energy* **2014**, *68*, 515–523. [CrossRef]
87. Sarasúa, J.; Martínez-Lucas, G.; Platero, C.; Sánchez-Fernández, J. Dual Frequency Regulation in Pumping Mode in a Wind–Hydro Isolated System. *Energies* **2018**, *11*, 2865. [CrossRef]
88. Sarasúa, J.I.; Martínez-Lucas, G.; Lafoz, M. Analysis of alternative frequency control schemes for increasing renewable energy penetration in El Hierro Island power system. *Int. J. Electr. Power Energy Syst.* **2019**, *113*, 807–823. [CrossRef]
89. Rezkallah, M.; Chandra, A.; Singh, B.; Singh, S. Microgrid: Configurations, Control and Applications. *IEEE Trans. Smart Grid* **2019**, *10*, 1290–1302. [CrossRef]
90. Yeager, K.E.; Willis, J.R. Modelling of emergency diesel generators in an 800 Megawatt nuclear power plant. *IEEE Trans. Energy Convers.* **1994**, *8*, 433–441. [CrossRef]

www.ingramcontent.com/pod-product-compliance
Ingram Content Group UK Ltd.
Pitfield, Milton Keynes, MK11 3LW, UK
UKHW060124010826
14089UKWH00006B/144

9 783036 538310

MDPI
St. Alban-Anlage 66
4052 Basel
Switzerland
Tel. +41 61 683 77 34
Fax +41 61 302 89 18
www.mdpi.com

Energies Editorial Office
E-mail: energies@mdpi.com
www.mdpi.com/journal/energies